Dr. Johnson's use of historical anecdotes, scholarly facts of evidence and pure reason correctly call into question our culture's fractured mores, which as he rightly concludes are built on sinking sand. These broken faculties of our society can only be repaired by returning to the values of Truth, a strong work ethic, and the underlying principles of capitalistic free enterprise.
– Bruce Powell, Senior Pastor/Teacher,
Siler Presbyterian Church, Matthews, NC

America has produced more freedom, wealth, innovation, and human opportunity than any nation in history, yet its founding Christian principles are under siege on every front. In Robin Johnson's words, "America isn't great because God made it more special, it's because Americans [at least in the past] made God more special." Dr. Johnson's profoundly insightful book for our turbulent times does not just diagnose the disease spreading through the soul of our country but provides solutions. The 21st Century Handbook should be required reading for every citizen in America as it applies historical and biblical perspectives to America's rapidly accelerating immorality and mindboggling social/intellectual confusion.
– Bruce Malone, retired Research Leader for Dow Chemical
and Founder of Search for the Truth Ministries

If you are one of the many that feel the Great Flood, Noah, and the description in Genesis is just a nice story, but evolution is how we got here, then you must read this compelling book on America's baffling culture war. The real history and science, along with the fascinating examples and details that Dr. Johnson presents will change your mind. Not only was the Flood real, once read you'll discover that it's logical and the only legitimate option as to our origin.
– David E. Samuel, PhD, Wildlife Biology (retd), West Virginia University;
book author, magazine and newspaper columnist

Welcome to the battlefield where each man is on the front line and their "manhood" is being tested daily. Are you worthy? Can you handle the attack? Will you prevail as an honorable man? Dr. Robin Johnson's 21st Century Handbook will take you on a journey to discover true manhood, the culture wars we are trapped in, and our universal need to consider pursuit of a more honorable life. Topics are broad-based, challenging, and sometimes difficult to accept—I am still working on several. It is through this struggle that we can all become more honorable—to all of mankind and to our Lord."
– Vann Walters, Technology Sales Executive, church elder,
father of daughters who deserve good men

The 21st Century Handbook is an in depth look at history, philosophy, science, and social issues that affect and shape our society and worldview. It is a well-researched and insightful book that calls for deep concentration and contemplation. If you're looking for a project that will challenge you mentally and spiritually, you'll find it here.
— Dr. Steve Jirgal, Founder and Director of the Jirgal Leadership Institute;
D. Min., M. Div., M.A. Health Ed., M.A. Sports Medicine

Enjoy a veritable cornucopia of philosophy, history, real science, and religion creating a road map through our Culture War for life success, with Jesus and Christianity taking the lead in the center lane."
— Dr. Keith Sellers, D.D.S., M.S. Orthodontics,
Sellers Orthodontics of Charlotte, N.C.

THE 21ST CENTURY HANDBOOK:
Cultural Chaos, Real Men, DNA, and Dragons

Robin D. Johnson

LIBERTY HILL PUBLISHING

Liberty Hill Press
2301 Lucien Way #415
Maitland, FL 32751
407.339.4217
www.libertyhillpublishing.com

Unless otherwise indicated, Scripture quotations taken from the King
James Version (KJV) – *public domain.*

Printed in the United States of America.

Paperback ISBN-13: 978-1-6312-9224-8
Dust Jacket ISBN-13: 978-1-6312-9225-5
Ebook ISBN-13: 978-1-6312-9226-2

For my two fine sons,
and confused truth-seekers across the fruited plain

CONTENTS

* Note: Other than when used to delineate an article or book title, *italics* are the author's and are used for emphasis, as is occasional use of bold highlighting. Quotation marks, when used on short phrases but not foot-noted, indicate being taken from the King James Version (or Authorized Version) of the Bible, unless otherwise indicated. The conventional, his-torical, and proper use of gender-indicating pronouns is practiced within. When referring to the more secular and liberal American political party, whose symbol is a donkey and is the party generally endorsed by the mainstream media, higher academia, the National Education Association, the American Bar Association, the American Psychiatric Association, the National Organization for Woman and the Radical Feminist movement, Homosexual Activists and Transgender movement, the American Civil Liberties Union, Hollywood, the victimhood and identity politics pushers as well as so-called social justice warriors , the term *Democrat* is used, instead of the more frequently used (but now inappropriate) term, Democratic.

PREFACE

BORROWING A SHORT phrase from the incomparable Theodore Roosevelt, I have been "in the arena" long enough to realize that no matter how factual, logical, well-reasoned or truthful my statements, criticisms and viewpoints might be, there are those who will disagree. I understand that. It's kind of like Ford . . . or Chevrolet? Some ol' boys out here in fly-over country who drive a Ford truck (like their dad and granddad did) will never buy and drive a Chevy, and vice versa-you're not going to change their mind. On this and particularly more important issues, many people are intransigent and simply "willingly are ignorant." Those who disagree, holler, or have a conniption fit because of something I discuss, point out or claim in this book are sincerely encouraged to just hang in there, check the facts as well as the fruit, and enjoy your quest for truth, as I finally did. I agree with German philosopher Arthur Schopenhauer on only one point: "All truth passes through three stages. First, it is ridiculed. Second, it is violently opposed. Third, it is accepted as being self-evident."

If you are a snowflake, mushhead millennial or coffeehouse misanthrope that can't think for yourself, this book probably can't help you much. But if you're not educated or propagandized out of the brain you were born with, have some backbone and can keep an open mind, please consider just checking the facts and real evidence, but also becoming a good fruit inspector. In life, when one is seeking sound medical advice, considering marriage, starting a new business, or even wants to learn how to play good golf, he or she should "check the fruit on the tree" of any particular individual before embracing their guidance or advice. The aforementioned Roosevelt said the person "Must evidence command of character", and "Look then to the works of his hands. Hear the words of his mouth. *By his fruit* you shall know him."[1] Sound advice indeed, and cannot this also apply to larger entities and issues, like nations, governments, social and economic policies, as well as even religions or gods? What's true and legitimate? What works best? What is *real*? Part of the answer: look at historical results-check the fruit on the tree.

Although there is indeed now serious cause for concern in America because of oppressive political correctness, bias and censorship in academia

---

[1] David Johnson, *Theodore Roosevelt: American Monarch*, American History Sources, Philadelphia, 1981, p. 276

and the mainstream media, herd-think promotion and the emerging *thought police*, in most quarters -but not at Cal-Berkeley, Louisville, Rutgers, Yale, Harvard and their ilk- one still has freedom of speech and thought. One caveat though, in the words of G. K. Chesterton: "You have the right to believe anything you want, but the truth is still the truth."

Robin Johnson, North Carolina flyover country, 2019

ACKNOWLEDGEMENTS

SPECIAL THANKS ARE due to my beautiful wife Janet, who has supported me in this book project and been an incredible blessing in my life for well over three decades, while offering her timely and sometimes profound wife wisdom, or woman's intuition, which I was somewhat ignorant of at the time of our marriage covenant. Dr. Henry Morris, often considered the founder or catalyst for the modern Creation Science movement, as well as the founder of the Institute for Creation Research (ICR), was a brilliant steel-backboned historian, author, engineer and scientist, Bible scholar and committed truth-seeker second to none, and continues to be an inspirational model to me, as does his ICR co-founder, former biochemist and science/origins debater nonpareil Dr. Duane Gish. To Dr. Harry Reeder goes great appreciation for his insight and teaching on real history, Biblical Truth, and true Christian manhood, including his unique insights on character and command in times of war. I will always remember Dr. Manfred O. Meitzen for first sharing the concept that *culture is religion externalized* with this sluggish undergraduate student so long ago, much later creatively emphasized by talented author and teacher Dr. George Grant (at Covenant Classical School, Concord, N.C.). These men inspired more commitment on my part to "redeem the time," and a more serious effort to live more from a Galatians 6:9 perspective. Thanks also to Wayne in my dental school days, for his counterbalancing conservative worldview, along with Dr. Dave Samuel, who helped me fall in love with a recurve bow, and became a great friend for a confused young man. Final thanks go to Nathan and Creed, my two fine sons, who inspired this book, and continue to stand for truth in a confused culture.

See Closing on page 309

INTRODUCTION

I LOVE TO OCCASIONALLY mosey through bookstores but am amusingly perplexed during such strolls by certain books that are there, and the subject matter observed that surprisingly even merited a book. But then again, "of making many books there is no end; and much study is a weariness to the flesh" (Ecclesiastes 12:12). Wow-too many books- written even three thousand years ago! Surprise may be unwarranted, though; verse 1:9 in Solomon's same book tells us "There is no new thing under the sun." So why another book?

Years ago I was relaxing with my two sons, then about 7 and 9, half watching a ball game on television when my first experience with a Viagra commercial on erectile dysfunction began to roll. I was stunned, then uneasy-as my sons focused with furrowed brow on the screen- and then I quickly became downright irritated. My mind flashed back to a popular TV sit-com being admonished and fined for just jokingly mentioning hemorrhoids on an episode when I was a kid many years before. What a change! Leftist media and academic types, political correctness, D.C. Beltway elites, the advertising industry, and the escalating *culture war* had me a bit bothered already. The genesis of this book was simply the subsequent search for answers related to the frustrating and puzzling changes in American culture, and all of the *mis*information out there on so many fronts, including the emerging new but blurred roles and definitions of American manhood. Initially I simply wanted to put together a little cultural guidebook for my two boys and their buddies. I procrastinated, and since that uncomfortable day on the couch with my sons years ago, things have gotten much worse, with not only entrenched and stifling political correctness (such an obnoxious term) but the rise of the victimhood and apology culture, identity politics, deep state government corruption, so-called gender *fluidity* and *dysphoria,* radical feminism, a pornography explosion, horrific school and church shootings, and an almost unfathomable expansion of power and influence by Big Tech and social media. A dishonest mainstream news media with an un-camouflaged secular, left-leaning political agenda—that has lost the public's confidence—is now obvious and undeniable. Our students experience a sad and slanted dumbing-down of American education, including revisionist history along with censorship and dishonesty in science teaching.

Unbelievable to many Americans, we also now have promotion of full-term abortion (infanticide), amazingly spearheaded by pediatric neurologist and current Virginia Democrat Governor Ralph Northam (whatever happened to *primum non nocere?*) and fellow Democrat Governor Andrew Cuomo of New York. Unbeknownst to many citizens, colleges and universities are now dominated by ungrounded *Women's Studies* departments with the concomitant suppression and vitriolic lambasting of the American male and fatherhood. Even pop singer Taylor Swift is now into some serious man-bashing in her newest music video (called 'The Man'). Young and millennial men now don't know how to act, often don't know what to say, and haven't a clue about real manhood. They need some truth-telling.

Additionally, the United States populace has become the unhealthiest among all Western and industrialized nations, despite having the most expensive and high-tech medical system in the world. Obesity is widespread-now 42% of Americans- with refined sugar, depleted wheat, and *fast* and *processed* non-foods reigning supreme. We have a serious prescription drug epidemic, as well as an ongoing illegal drug dilemma. The U. S. has nearly 5% of the world's population but unbelievably uses about 60% of the world's prescription drugs. The sage Samuel Johnson said many years ago: "Health is so necessary to all the duties, as well as the pleasures of life, that the crime of squandering it is equal to the folly." Well spoken, indeed. Excessive and unjustified medical testing, the overuse of scans, over-diagnosis and overtreatment, and resistant virulent strains of bacteria-mainly due to the overuse and abuse of antibiotics- have all become serious problems in American health care. Misinformation abounds here as well. Health and wellness truth will be lightly touched on.

I work in the health-care field but do some teaching as well. Some of my students probably get tired of my maxims and clichés. One overused favorite: *Culture is Religion Externalized*. This one was borrowed from a professor who taught a Religious Studies evening class taken as an elective while I was attending dental school years ago. I'll never forget him-he emphasized critical thinking-rare in academia today. He was brilliant and challenging, chairman of the department, and to my surprise somewhat conservative. He often wore cowboy boots, was patriotic, and loved to bear hunt— a real anomaly in the underworked, overpaid, college professor crowd. His point: what we see around us in the culture-drug use and crime rates, trends in the arts, language changes, sexual behavior and mores, public discourse and manners, and even things like hygiene and public health or architectural styles are all related to that culture's religion and moral compass, or the lack thereof.

I should point out that I was not a Christian then, and rarely went to any church. The main reason: my belief that so-called science had disproved the Bible, especially its foundational historical narrative from the early chapters of Genesis, and that evolution and billions of years were reality. In actuality I was, like so many of you, unknowingly a victim of censorship and *white lab coat intimidation*.

Shifting gears for a moment, I had by that time a lot of experience as a scuba diver and spear fisherman, and after much time on shipwrecks and the ocean bottom came to realize that what I'd been taught in school, textbooks, and museums about paleontology and fossil formation by *the white lab coats* (science teachers and professors) was totally false. In other words, the scenario of an animal dying in a marine environment-whether large fish, whale, land mammal, bird, or even dinosaur-and sinking slowly to the ocean floor, laying there undisturbed for many hundreds or thousands of years, while slowly and gradually being covered up by sand and sediment and eventually turning into a stone-like fossil, was an absolute *impossibility*. Why? I realized and learned, not from college textbooks and professors, but from real experience-what I call *evidence-based* science-that most dead animals in water (including humans) tend to bloat and float and are quickly scarfed up and eaten by scavengers and predators, broken down by bacteria, other microbes and worms, or eventually ripped asunder by wind and wave action.

I had been fascinated with dinosaurs and fossils since a youngster, my first treasured book being *All About Dinosaurs,* by Roy Chapman Andrews. But this realization struck me, and stuck with me as extrapolated estimates based on more and more worldwide fossil finds pointed to millions—if not even *trillions*—of dead things (fossils) in the vast sedimentary rock layers (those laid down by water) spread all over the earth. In today's world, fossil formation is an extremely difficult and very rare process; it takes unique and special conditions, and is seldom observed happening anywhere. For example, out where I used to bow-hunt in Oklahoma and Kansas, and across much of the Midwest prairies, multiple tens of millions-some estimate well over a hundred million-buffalo (bison) roamed the American plains just 150-175 years ago, but we have never found a single undisputed buffalo fossil. Conversely, we now have evidence indicating uncountable numbers of *marine* fossils-billions-even on the top of all of the tallest mountain ranges in the world, underneath the snow and ice, on the high summits of the Rockies, Andes, and even the Himalayas of Asia, which include many peaks over 25,000 ft. tall. How did they get there?

Having always been fascinated by the natural world and how things came to be, I started to wonder, embarrassingly at first, if maybe . . . just maybe . . . there had been some sort of catastrophic worldwide flood in the past. The massive thickness and sometimes *continental size* of the water-deposited sedimentary rock layers, as well as the nature, configuration, and stunning number of buried life forms precluded any preposterous *local* flood explanation. And, what about that Noah dude of biblical renown? Was he real? Could there possibly, really, have been an actual global Flood of Noah's day? After all, it had become apparent to me that quick, massive, violent inundation of sediment on a prodigious scale—*fast and furious*—not slow and gradual, was the only scenario that made any scientific sense. This started my sluggish quest, and this is why—although it may seem silly to you readers—I enrolled in that biblical studies night course while in dental school. I was lookin' for answers.

I was transfixed when the professor led a few classes on "The Divine Imperative for Ethics," and was stirred within even while reluctantly realizing I was about as sharp as a marble. Quite some time after the course was over, it finally sunk in that if evolution and billions of years were true, and we were just an accident of random cosmic happenstance, then we really have no legitimate basis for any ethics or morality; there cannot be justifiable moral absolutes, and consequently no moral authority from the Bible (or any other religious text) as well. Can life then truly have *any* meaning?

What idea or concept could be more important to not just teenagers, or millennials, but to anyone, than how they view their origins? Think about it; optional *eternal* ramifications might be at stake. As a young man I was steeped in evolution and long ages, had a college science degree, was not a church-goer, had a questionable lifestyle, and was focused on self. But my assumptions, and my comfort zone were starting to crack, and my view of the world was beginning to crumble. How about you young adults, millennials, or even *baby boomers* (like me)-do you have a guiding and grounded *worldview*? Or do *you* just go with the flow?

This book is not for the low-information voter or the Ph.D. in the rarified air behind the safe ivy-covered walls of higher academia. This *Handbook* is for the people across our country that work hard and make this great nation *hum*, and who still treasure fairness, family, and faith and love America. But they sense something is terribly wrong, and wonder not only *how we got in the mess we're in*, but what to do about it. *This 21st Century Handbook* reveals and explains. It first shares heretofore suppressed and censored truth, and then some needed actions and remedies, specifically directed at those who desire to save American exceptionalism.

The wimpification and assault on American men is directly related to the accelerated leftist leaning in higher education, the power of radical feminism, stifling political correctness, and a much-touted and taught but ungrounded starting assumption and worldview. Let's look honestly at what is really going on. I will discuss the initial wrong turn by higher education, so-called *toxic* masculinity, and the foundation and characteristics of real American manhood. Profound misinformation on the Confederate Generals and monuments issue, as well as the attack on our Founding Fathers will be covered. Revisionist history teaching and widespread academic pseudoscience storytelling—contrasted with *real* science—particularly as related to the origins issue, geology, genetics, and dinosaurs will be discussed and exposed. The *why* of America's horrific mass shootings and the unprecedented power and influence of Big Tech/Silicon Valley will be discussed and exposed. A serious challenge is issued against the foundational assumptions and unfortunate changes in our educational and legal systems, which have contributed to countless confused young adults and societal chaos. Are you worn out from reading that? It's even more wearisome trying to figure out how it happened, and what to do about it . . . on your own. The hard-working folks need a li'l help; they need a *Handbook*. Believe it or not all of these issues are related, because as you will discover, culture *is indeed* religion externalized.

PART I

MANHOOD AND MISINFORMATION

1

MEN AND WOMEN, PSYCH DOCS AND THE MOCKINGBIRD

WE ARE NOW immersed in an exasperating and very serious culture war in the United States, more and more frequently referred to as a *cold civil war*. The reeling recipients from the point of the spear by the widespread assault of political correctness, social justice warriors, leftist academics and particularly their Women's Studies departments are American men. Men-particularly young men and white men-are confused, intimidated, told they're responsible for most of the world's problems, often don't know what to say, how to act or react, and haven't the foggiest idea about real authentic manhood. They have been misled, lied to, repeatedly taught *white privilege* while at the same time indoctrinated—along with other students—they all evolved from pond scum. They've become the primary victims of revisionist history and misinformation by academic elites, and slammed in other ways. Young African-American men and women and other ethnics have been sent a message by the liberal left, the mainstream media, identity politics pushers and the Democrat political party that they need special favors, affirmative action, relaxed standards, and now even an *adversity score* factored in to improve their SAT performance. Many proponents of these policies do not realize how racist this mindset is. The idea that these particular young people are not quite up to snuff—the subliminal message that they're not quite as smart, that they need extra help and favors; that, well, they just can't usually make it on their own— is sad, false, and hearkens back to the early 1900s when this view was sometimes taught in our museums and schools. Such a view is evolution-based, racist, un-American, counterproductive and self-image assaulting. It *is* encouraging to finally see more sharp people of all colors and ethnic groups starting to recognize what is really going on, and the unfortunate message and political agenda being promoted.

Surveys indicate women now make up some 66% of college students, quite a significant change from when I was an undergraduate student forty years ago, as is the quick rise of questionable, manhood-attacking

3

Women's Studies departments. Is it any wonder that young men, more and more, are staying away from the hallowed halls of academia; the leftist politically correct indoctrination and abuse starts as soon as they set foot on campus. Some went to private Christian high schools, home schools, or regular public high schools with still at least a modicum of common sense and normalcy, but these young men are a bit stunned to find out quickly that *they* are the real societal problem, particularly if they're Caucasian. They are inundated with often false Women's Studies hogwash, LGBT studies and promotion, gender fluidity acceptance, *tolerance* of every-thing, and much more. For the uninformed, there have even been pro-posals by some feminists and women's studies advocates from academia to do away with the word *history,* and replace it with the term *herstory* -*her* story-they claim it really is women and their deeds that have unques-tionably been the important and dominating influence in the unfolding of history and creation of culture throughout the ages. Many Americans have no idea of the dominance of radical feminism in the political sphere and academia, and their role in what political scientist Stephen Baskerville calls the *New Politics of Sex.*[2] In referring to the various leftist ideologies that have found their stronghold in higher academia, he refers to the opinion of highly respected former federal judge and Solicitor General, the late Robert Bork: "feminism is by far the strongest and most imperialistic, its influence suffusing the most traditional academic departments and uni-versity administrations."[3] It's no wonder that many young men now exhibit timidity and lack of confidence. Many hesitate at even politely opening a door for a female, fearful of a gender neutral/politically incorrect backlash or tongue lashing.

Another obvious major focal point by the secular left and higher aca-demia is to not only minimize, but *eliminate* any and all gender differences in our society. In some academic institutions, feminist groups, and county/ city governments, there has even been a successful effort to eliminate the use of gender pronouns like *he, his*, and *she*, and sports terms may now even be placed on the chopping block. Like in baseball, the first base*man* or second base*man*-for you unenlightened slugs in flyover country (like me)-first and second base*person* is preferred. It would be downright funny if history didn't show that the downfall of every great nation-state is *preceded by a change in the language*; it's been happening here at an

---

[2] Stephen Baskerville, *The New Politics of Sex*-The Sexual Revolution, Civil Liberties, & the Growth of Governmental Power, Angelico Press, Kettering, OH, 2017

[3] Ibid, p. 3, cited from *Slouching Towards Gomorrah*, Ch. 11, 193-225,excerpte at http://fathers for life.org/feminism/borkch11,htm

accelerated pace in recent decades. As usual, liberal Berkeley, California, the city that birthed free love, the hippie movement, and the modern psychedelic drug culture has helped lead the way (and incidentally, just banned-late 2019- the use of natural gas in new construction). We have indeed some new, heretofore non-existent issues to deal with, and many citizens feel it's gotten a little crazy out there. Are these new ideas sound, and hold water? Where did they come from, and might they be beneficial to society? Or are we simply losing our collective minds?

Although apparently at odds with the view of some progressives and all gender fluidity proponents, the most obvious difference between men and women is biological. Be careful and circumspect around anyone who denies this. The brilliant pediatric neurosurgeon and current head of Health and Human Services in the Trump administration, Dr. Ben Carson, who reeks with knowledge and credibility about health and the human body, was recently criticized by the left for saying there *are two* genders! Those that do actually deny this, and others that at least deny the unique gender abilities and distinctions of the two sexes—which science, history and social interaction can confirm—claim that these are *not* real, innate, or programmed (in the DNA) differences; they are just *forced* on us by society. I have a friend who conversely contends that men and women are not only good at different things, but are just *wired* differently (I tend to agree), and he offers numerous interesting and entertaining evidences of that claim. Insightful and appropriately humorous columnist Andree Seu Peterson, in a recent *World Magazine* commentary, chimed in with her always interesting perspective. She more than hints at gender distinctions and differences, and why she's *thankful* for men:

- They're stronger than women; they can lift more weight.
- They don't complain about lifting those heavy things for women (they rather seem to enjoy it).
- Men die by the millions in wars defending women, and don't complain about that either. [They also don't participate] in street demonstrations . . . protesting the lack of women in combat. [Sometimes men are even seen] marching in women's marches, even when those women are marching against men. One cannot imagine it the other way around.
- Men do most of the inventing of neat stuff that exists in the world, either for international defense or for comfort. [Even] the techy gadgets women use to broadcast hate for men were mostly made by men.

5

- The greeting cards at pharmacies have jokes at the expense of men, and never vice versa. They make the wife look smart and the husband look stupid.
- Men are the best music composers.
- Men are the best artists.
- Men are the best writers.
- Men are the best chefs.
- Men are more courageous than women. When it sounds like someone is breaking into our house in the middle of the night, it's my husband who goes downstairs with the baseball bat, not me.
- Men brave Arctic cold on deep sea rigs to pump oil out of the ground to warm our homes . . . [and] coax coal from stubborn veins embedded deep in the earth.
- Men say cooler things to each other in private.
- Men cut to the chase (except William Faulkner) . . . (*I agree; but Edward Everett is right up there with Faulkner-RJ)
- Men are better at changing flat tires.[4]

It is so satisfying to hear or read straight talk that makes you smile. Check out *WORLD* magazine-it is excellent.

More and more of those on the political left and in the radical feminist movement deny such inherent differences, partially catalyzed in recent decades by secular psychologists and white lab coat *behavior experts*, who claim that all such perceived psychological, aptitudinal, or intuitive differences between men and women are due to a *required* or *forced* socialization. This odd claim goes as follows: apparent differences in predispositions, proclivities, interests and abilities between the sexes are *acquired*, or required—*socially enforced*—not inherent, naturally embedded, or created and programmed by a Creator God. (This somewhat calls to mind the disproven and embarrassing claims—to evolutionists—of Lamarck and Charles Darwin's of the inheritance of *acquired* biological and morphological traits, long proven false, after duping untold numbers of innocent students). To restate this novel, even radical, male-female/no differences in-the-sexes idea—and this is puzzling new ground for me as well—the heretofore easily recognized, worldwide acceptance and general appreciation of gender distinctions is now, well . . . *fake*—they're not really real— they are just *acquired*.

This cultural and psychological gobbledygook originates in part from the American Psychological Association and their ilk, the same group that

---

[4] *Thank God for men*, Andree Seu Peterson, WORLD Magazine, Nov. 11, 2018

in the 1970s changed its characterization of homosexuality (formerly sodomy) from a *mental disorder* to just *an alternative lifestyle*. This idea of *forced* socialization, from the gay/transgender lobby's lap-dog psych group, in our sudden gender dysphoric culture is not fascinating, insightful or valid—it is preposterous. It is highly recommended that you take any proclamations or psychobabble from *behavioral* scientists, psychiatrists or psychologists with more than a few grains of salt. I can remember well a biology/animal behavior self-styled expert and professor I had decades ago who claimed to us students that virtually all animal behavior is *learned*, or acquired. As a then 20-year-old who grew up in the hills of West Virginia enjoying the outdoors, and even back then observing and being fascinated by the miracles of animal and bird migration, their nest-building and song patterns, fish spawning, and mammal mating seasons and habits—just to name a few amazing examples of built-in, or *instinctive* behavior—I thought, *What universe did this stuffy dude grow up in*?! Of course all of those behaviors are **not learned** but are amazingly **programmed**, or coded, in the animal's DNA. A simple example would be when a songbird finally is strong enough and mature enough to leave the nest and go do its own thing, find a mate, and build its own nest like, say, a mockingbird. No person, no bird, no computer, nor even Google teaches them *how* to build that nest. The young mockingbird was never taught how to build a nest, yet eventually off it goes, and guess what? It not only successfully builds a nest, but it builds a *mockingbird* nest; not like a pine warbler's nest 20 ft. or higher hidden in pine or spruce needles near the tip of a horizontal branch, nor like a meadowlark's partly-domed nest well-hidden in the grass on the ground in a hayfield, or the Carolina wren's unique little nest in an old oil funnel in my barn tool room (once in an old tractor engine). The mockingbird nest is always about 5-15 ft. off the ground in a big bush or small to medium-sized tree; the nest is always the *same* size and circular shape, easily identifiable as a mockingbird nest. Many of the passerine (or perching) songbirds migrate (head south) for the winter, then head back north in the spring (another example of embedded, programed information), often to the very *same* geographical area (sometimes the exact *same* spot), and always build the *same* kind of nest as their parents did, or as they themselves did the previous season, in the *same* kinds of places or trees, about the *same* height off the ground, with the *same* type of materials. The bird was never taught how to do this, but *it knows*; it is programmed information, just like a bird's characteristic song (not *acquired* or learned). It is important to understand that information, and especially *coded* information (think DNA), never happens by chance, or originates from randomness; this will become more apparent to the reader later in this book. The building of a bird nest is a phenomenal event, as is construction of even those ubiquitous round or oval-shaped leafy squirrel nests up in the trees. Consider not only *how* do they know what to do (they were never taught), but who holds the first few sticks or leaves while they skedaddle off to gather more building materials? Have you ever wondered about that? I sometimes think our theological superficiality and weakness, and relegation of spiritual issues to the back-burner, comes from our failure to detect the obvious. I know urban city folk don't always think along the lines of a hillbilly like me, but nests are such cool little miracles of *programming*, whether you're a Christian believer or not. It's really amazing—without nails, screws and mortar—those nests usually don't even go down in bad storms! A lot can be learned from the critters. I even agree with atheist philosopher/mathematician Bertrand Russell on one point, where he said "the world is full of magical things patiently waiting for our wits to grow sharper." Russell, though, felt those amazing, magical things somehow *evolved* out of the primeval slime through random mutations, which has **never** been observed (he reminds me of Romans 1:22). The biblical Job, nearly 4,000 years ago, suggested: "ask now the beasts, and they shall teach thee; and the fowls of the air, and they shall tell thee . . . and the fishes of the sea shall declare unto thee." Dante, 700 years ago, said it well: "Nature is the art of God." Following that thread, the Bible actually claims that God has given numerous evidences, or *reasons, to believe*. Here is one, from Romans 1:20:

> For the invisible things of him from the creation of the world are clearly seen, being understood by the things that are made, even his eternal power and Godhead; so that they [unbelievers] are without excuse.

When I read that part about the *invisible* things I think not only of the many examples of programmed behavior and coded information in animals and plants, but things like gravity, electricity and electrons, magnetism and their known predictable interactions. Also Maxwell's and other mathematical equations, the laws of science, and the profound order and balance in the world, what even ardent atheistic evolutionist Richard Dawkins reluctantly described as the "appearance of design" everywhere. Most of these amazing phenomena, in addition to animal migration, mating behaviors, DNA-directed differentiation and growth, and so much more in the panoply of the earth's organisms and processes are based on *information*. Information **never** comes from randomness (which the evolution story is based on), including the mockingbird's micro-embedded unseen nest-building programming. The divinely designed unique songs of other beautiful bird species are additional *pro-gramming* examples, whether it's the "drink- your- TEA!" of the towhee, the "TEA-kettle, TEA-kettle, TEA-kettle, tea" of the Carolina Wren, or the emphatic ending of the chestnut-sided warbler, "very very pleased to MEET-CHA!"

Significantly, Christianity is the only major world religion that even talks about creation, or beginnings (where we came from), and is the only one actually *set* in history. That is, many of the Bible's protagonists, stories, events, claims, and prophecies can be checked and verified by extra-bib-lical sources, historical writings and ancient secular records; documenta-tion and confirmation of biblical events; as well as scientific, geological and archaeological findings. **No other** religion is like that. It still requires faith, but it is not a *blind* faith—like that of other religions—it is grounded in history. This is not proof, but it *is evidence*. Don't take my word for it, and don't be intellectually lazy-check it all out.

Let's go back to the psychologists, *acquired* traits, and their question-able claims and history. The past century's field of psychology was based much on the musings of a cigar chain-smoking Austrian secular evolu-tionist, Sigmund Freud, who despised Christianity and was fixated on *libido* (sexual desire), but whose ideas have now largely been refuted. Freud considered Christians ignorant, and their foundational belief "so patently infantile, so foreign to reality [as well as] not tenable [and one that] con-sists in depressing the value of life"[5] (it doesn't sound like Dr. Freud ever carefully read the Book, or studied history). Others that followed Freud also operated from an evolutionary foundation, like Carl Jung, Carl Rogers, J. B. Watson, B. F. Skinner (with his ungrounded *self-esteem* theory) and a large majority of the more recent ones who seemed to despise Christianity, like Harvard's Edward Wilson, or Wendell Waters, former professor of psy-chiatry at McMaster University, who said this:

> I want you to entertain the hypothesis that Christian doctrine, the existential soother par excellence, is incompatible with the principles of sound mental health and contributes more to the genesis of human suffering that to its alleviation. . . In my view, all religions are inhuman anachronisms, but here I am only dealing

---

[5] Sigmund Freud, *Civilization and Its Discontents*, (Translation from the German by James Strachey), W.W. Norton & Company, Inc., New York, 1961, pp. 21, 31.

> with Christianity and, more specifically, with the noxious nature
> of Christian doctrine at the personal and interpersonal levels.[6]

"*Noxious* nature"!? This is just stunning, especially the professor's ignorance of history, but even more so the reality of many people, especially Canadians, coughing up cash to pay this dude to teach their kids. It's too bad the Bible-hating psychiatrist never read historian Tom Holland's *Dominion: How the Christian Revolution Remade the World* (2019), as Mr. Waters passed away in 2012. It is really not surprising though, as many of the modern-day psychologists consider religious belief—especially Christianity— and especially those who believe in a recent Creation and the Genesis historical record, as characteristic of those having *mental* problems. These psychiatrist-psychologists' common denominator, or shared foundational precept: belief in evolution and billions of years, with man just another evolved animal.

When I was first exposed to some of these guys, it was just disbelief with slow head-shaking, as I wondered how they even came up with some of these ideas. . . like those of socio-biologist Edward Wilson of Harvard, once considered the world's leading authority on ants. With his religious faith in evolution, and hatred of the Church, Wilson said this (referring mainly to Christianity):

> Bitter experience has taught us that fundamentalist religion
> . . . in its aggressive form is one of the unmitigated evils of
> the world.[7]

Subsequent comments and writings indicate that Wilson (who is still alive; will be 91 in June 2020) directed his wrath particularly against Christianity. I have occasionally wondered, when stumbling upon Wilson's dogma, how the great Christian scholars and professors from Harvard's first 230 golden years would feel about his worldview and teaching content. Wilson and coauthor/philosopher Michael Ruse, along with many other evolutionists, believe that even man's ethics and moral inclinations (as well as our physical and morphological structure) are merely the product

---

[6] Wendell W. Watters, "Christianity and Mental Health," The Humanist 47 (Nov./ Dec., 1987), p. 5

[7] Edward O. Wilson, "The Relation of Science to Theology," *Zygon* (Sept/Dec 1980); this controversial paper was presented at a conference sponsored by the Institute of Religion in an Age of Science, and the Academy of Arts and Sciences.

of undirected, random mutation-based evolution. Unsupported by *any* scientific evidence, here is their claim:

> Morality, or more strictly our belief in morality, is merely an adaptation put in place to further our reproductive ends . . . Ethics is seen to have a solid foundation, not in divine guidance, but in the shared qualities of human nature and the desperate need for reciprocity.[8]

Not only is this elitist comment unsupported by anything remotely resembling scientific evidence or evidence-based history, and qualifies as nothing more than evolutionary make-believe (real only in the authors' minds), but it is in opposition to the Bible (e.g., Romans 2.14-15), as well as the prevailing opinion not only of most of our Founding Fathers, but most of the greatest scientists since the Reformation.[9] Sadly, I discovered Wilson's books are used in several highly regarded private schools. Be circumspect about lending too much credence to psych docs and socio-biologists, as well as where you send your kids to secondary school and college. Incidentally, have you ever seen the statistics on how many psychiatrists have sexual relations with their patients? It is stunning. Ironically, they even seem to have a lot of personal and mental problems in their own lives as well. Peter Breggin, a psychiatrist himself, said this:

> The average psychiatrist has more power to do harm in the lives of individuals than most religious leaders on Earth. . . Moreover, it would be hard to find a more unhappy lot than those clustered in the mental health field. Especially among psychiatrists, suicide, depression, drug addiction and alcoholism are notoriously rife. Among non-medical mental health professionals, the situation doesn't seem much better. Not only are many mental health professionals unhappy, but

---

[8] Michael Ruse and Edward O. Wilson, "Evolution and Ethics," *New Scientist*, 208, Oct. 17, 1985, p. 50

[9] Henry M. Morris, *Men of God Men of Science-Great Scientists Who Believed the Bible*, Master books, El Cajon, CA, 1992 (Sixth printing)

they do not live ethically inspired lives. Too many, for example, prostrate themselves before the psychiatric establishment.[10]

People should be more skeptical in considering the claims of some-times bombastic white lab coat psychiatric experts, including the infamous Dr. Freud, who enthusiastically felt that the *libido* was everything and that cocaine was the answer to most of the world's problems (and cigar smoking, although throat cancer killed him). Another part of his legacy is the emphasis on *therapy,* and that *truth* is to be found mostly in a dia-logue with the *unconscious*. It is sadly ironic that the word *psyche* refers to the soul, and *psych*ology is thus the study of the soul. With psychology's foundational assumption of Darwinian evolution and man as just another evolved animal, these people *don't even believe in the soul*. Freud's con-tempt for God and worship of evolutionary dogma dominated the field of psychology; consequently psychologists assume that evolved humans' behavior and psychological problems must be evaluated from that animal-istic foundational starting point, or assumption. Logically following such a worldview or underlying presupposition is non-belief in any moral abso-lutes. But what if their starting assumption, or *primary axiom,*[11] is *wrong*? We'll delve into that throughout this book.

This attempted denial and elimination of gender differences, and therefore also of different male and female roles, is everywhere today, reaching a crescendo and literally bursting on the scene in the last couple of decades, to the bewildering surprise of many hard-working Americans, especially those of faith. As Robertson McQuilkin pointed out some years ago in *Biblical Ethics*:

> The "unisex" view of human nature recognizes no inherent distinctives apart from the basic physical distinction that all must grant, however reluctantly. Most who advocate the unisex viewpoint are strongly in favor of eliminating all role distinctions. Roles based directly on the biological functions necessary to fathering and mothering (perhaps "inseminating" and "bearing" would be more accurate descriptions) are accepted,

---

[10] Peter R. Breggin, "Mental Health versus Religion," *The Humanist* 47 (Nov./Dec. 1987): p. 13

[11] Dr. John Sanford, *Genetic Entropy*, Prologue, p. vi, and pp. 5-9, FMS Publications, 2014

but all other role distinctions are negotiable, dispensable, and may, in fact, be pernicious, according to this view.[12]

This is very much opposed to the biblical view, where *equality* of the sexes (without interchangeability) is stressed, but with different roles and responsibilities. Equality is simply one's standing; the same blood of Jesus was shed for a woman as for a man, and the same Holy Spirit indwells a Christian woman as indwells a male Christ-follower. Husbands are to treat their wives as "joint heirs." Biblically it is not a matter of superiority of inferiority; it is just a matter of *distinctiveness*. This scriptural view promotes other rights and non-favoritism as well: whether young or old, brown, black or white, wealthy or poor, highly-educated or little educated, the Christian God is "no respecter of persons." And the Bible **never** mentions or talks about *race*-only tribes and people groups. The modern-day use of and meaning of race is in large part an evolutionary construct; this will be discussed in detail later in this book. In Acts 17:26 Paul proclaims that God "hath made *of one blood* all nations of men." Back to the standard of **biblical gender equality; it clearly *does not* mean interchangeability. Decades ago, I heard a pastor** beautifully describe how men and women are *complimentary reflections of the image of God*. Today confusion, misinformation and strong disagreement currently reign on gender fluidity and gender *dysphoria*, terms that didn't even exist a few years ago. What has happened? What are the reasons, or history, behind the disregard and loss of biblical authority? And is there even such a thing as *real* men anymore?

---

[12] Robertson McQuilken, *Biblical Ethics*, Tyndale House Publishers , Wheaton, IL, 1995, p. 182

2

MANHOOD, ACADEMIC METAMORPHOSIS AND MATERIALISM

AMERICAN MEN AND manhood, military performance, leadership and the character of our Founding Fathers, as well as the characteristic American "can-do" attitude and spirit have, until recently, been recognized and admired at home and around the world. For several decades now, this view has *not* been the case with Hollywood and many pop culture celebrities, college academics, the mainstream media and the political left. What has happened? It is time for change and truth-telling, especially for our young people.

The secularization of American education—the elimination of any mention, teaching, basis, influence or reality of God or a higher power—is the foundation of this attack on men and our cultural metamorphosis. The rise and dominance of materialism, naturalism, atheism, and ungrounded and unfounded Darwinian evolution (they're closely related terms and pragmatically synonymous; take your pick) is apparent everywhere. Our once-great colleges were founded in large part as what I think of as *training grounds* for Christian warriors, or more specifically to train, equip and educate Christian pastors and leaders. Most moderns and current college students don't know this—not that it would change their worldview—but it is important, especially if we want to try to figure out *how we got in the mess we're in*. In the last century or more, these formerly outstanding Christian colleges and universities have transitioned, or more accurately *metamorphosed*, from a Christian worldview to a secular evolutionary-materialistic one. Or, put another way, from a Judeo-Christian consensus, to a humanistic-atheistic one. Decades ago theologian-historian Francis Schaeffer sadly described this drastic, culture-shocking change as moving from the reality of an infinite, personal Creator-God to a godless reality of just *material energy*. The idea: this material energy has supposedly existed forever, but has developed into its present form (including us humans) entirely by random chance. This assumption is refuted and challenged throughout this book (particularly in **Part II**, **Science and the Culture War**).

The word 'university' is a combination of the Latin *uni*, meaning to unify and *veritus*, which means truth, thus the main focus of these early institutions was to couple knowledge with biblical truth. This includes Harvard which was founded in 1636, William and Mary in 1693, Yale in 1701 and Princeton in 1746. Harvard has led the way in higher education since the early days. Its mottos were "For the Glory of Christ," and "For Christ and the Church,"[13] (laughable in Cambridge today) and the school's declared educational precept in 1636 was clear:

> Let every student be plainly instructed and earnestly pressed to considerwell the main end of his life and studies is to know God and Jesus Christ which iseternal life, (John 17:3), and therefore to lay Christ in the bottom as the onlyfoundation of all sound knowledge and learning . . . Every one shall so exercise himself in reading the Scriptures twice a day that he shall be ready to give such an account of his proficiency therein.[14]

Even over 150 years later in the 1790's, during George Washington's presidency, these guidelines and rules held sway at Harvard:

> All persons of what degree forever residing at the College, and all undergraduates . . . shall constantly and seasonably attend the worship of God in the chapel, morning and evening, . . . All the scholars shall, at sunset in the evening preceding the Lord's Day, layaside all their diversions and . . . it is enjoined upon every scholar carefully to apply himself to the duties on said day.[15]

---

[13] *The Harvard Graduates' Magazine,* George Barna Publishing Co. Manesh, WI, Sept. 1933, p. 8, from the article "Harvard Seals and Arms" by Samuel Eliot Morrison.

[14] Benjamin Pierce, *A History of Harvard University,* Brown, Shattuck, and Company, Cambridge, MA, 1833, Appendix, p. 5

[15] *The Laws of Harvard College*, Samuel Hall, Boston, 1790, p. 7-8

With today's extreme *tolerance* and *diversity* mindset and political correct, pluralistic culture, this is, of course, not required now and even sometimes mocked and ridiculed at Harvard and many other universities (and essentially banned from our public schools). I'm reminded of a joke I recently heard: How many Harvard students does it take to change a light bulb? The answer: Just one-he (or she) holds the light bulb, and the world revolves around them.

All joking aside, this former Harvard mission statement or foundational purpose produced some of the most amazing leaders not only in fledgling America, but anywhere in western civilization during that era. These men were instrumental in producing our unique system of representative government—what H. L. Mencken a century ago referred to as our "Great Republic"—still unmatched in world history, and despite the controversial and volatile challenges we're dealing with now, still going after almost 250 years. This is just remarkable, when compared to the *fruit* and track records of other governments and nation-states in recent centuries (see Part III). Remember, be a fruit inspector; what's good, what works? Where's the good fruit? Be honest. Have you objectively compared the United States to others, looking at the last 300 years? And how about those that are Buddhist, Muslim, Shinto, Hindi, animalist, socialist/communist or some other religion-base? By various parameters-whether individual liberty, the freedom to choose, GDP, economic opportunity, science and technology achievement, women's and human rights, or something as basic as clean water and modern sanitation availability for all—it's not even close. Don't misunderstand me—I am absolutely *not* insinuating that God made Americans greater or more special—the important point and difference is that Americans have *made God more special*. Great men and Christians John Hancock, Rufus King, John Adams, his cousin Samuel Adams and other Founders attended Harvard College.

Princeton University's first president, Jonathan Dickinson, boldly said this: "Cursed be all learning that is contrary to the cross of Christ." This New Jersey school produced more Founding Fathers than any other single college, including Benjamin Rush, James Madison, Richard Stockton and the Rev. Dr. John Witherspoon. The great Witherspoon, President of Princeton right before the Revolution, and leader of its revitalization afterwards, was a bold patriot and inspirational educator, and said this when leading Princeton:

> Every student shall attend worship service in the college hall morning and evening at the hours appointed and shall behave with gravity and reverenceduring the whole service.

> Every student shall attend public worship on the Sabbath .
> . . there shall be assigned to each class certain exercises for
> their religious instruction suited to the age and standing of
> the pupils . . . and no student belonging to any class shall
> neglect them.[16]

Wow, how things have changed at Princeton! Yale is one more example, perhaps the most egregious. In recent years I've wondered who the heck is even running the show at Yale, if anybody. Now it is obvious: its strident far-left undergraduate students. In its former halcyon days, Yale's educational philosophy/theology was openly Christian, with major emphasis on Biblical principles and prayer, with this particular requirement when classes began in 1701:

> [The] Scriptures. . . morning and evening [are} to be read by
> the students at thetimes of prayer in the school . . . studiously
> endeavor[ing] in the education of saidstudents to promote the
> power and purity of religion.[17]

And in 1720, another directive to the students: "Seeing God is the giver of all wisdom, every scholar, besides private or secret prayer . . . shall be present morning and evening at public prayer.[18] This is sadly ironic, considering the stifling blanket of atheistic groupthink, non-free speech for conservatives and Christians, sexuality focus, and political correctness engulfing Yale today. Founding Fathers and outstanding Christians William Livingston, William Samuel Johnson, and Noah Webster are just a few of Yale's great graduates.

Others schools likewise followed these three institutions, and in a good many of these early colleges one could not even be admitted without confession to belief in the Trinity and Christianity, sometimes both on paper and orally, in a serious face-to-face interview. The same applied in the early

---

[16] The Laws of the College of New-Jersey; Trenton: Isaac Collins, 1794, pp. 28-29

[17] *Documentary History of Yale University*, Franklin B. Dexter, editor, Yale University Press, New Haven, CT, 1916, p. 32; from Proceedings of the Trustees, 1701

[18] Daniel Dorchester, *Christianity in the United States*, Hunt and Eaton, New York, 1890, p. 245

days to state governorships (even written into some state Constitutions). The foundational precept was the truth of the Bible, with Jesus as Creator and Redeemer and divine lawgiver, and therefore the reality of life principles or moral absolutes. The idea essentially was (and still is among historical orthodox Christian believers) that the God of the Bible (and its history) is true and real. He made you and loves you, with his main focus to create and redeem (because of our rebellion, starting with Adam) men and women "in His image," through his Son, Jesus (implying tremendous worth and value to *every* human being, even in utero). Consequently, He owns us and sets the rules, for our benefit . . . thus, the presence of his statutes, commands, and precepts.

Sadly, the only absolute at Harvard, Yale and these other institutions today: *there are no absolutes*—everything is relative—so set your own rules, do your own thing. Half of the professors at these *elite* schools, that are supposed to be so outstanding, are *Marxist*; most even admit it. The atheistic far-left radicals and hippies of the 1960s became college professors and now run the universities, and are out to drastically change this nation. These liberal universities, like Rutgers, Brown, Columbia, most of the Big Ten and Big 12, even some ACC and Pac-12 schools (especially Cal-Berkeley), Yale and most Ivy League schools, and so many others have virtually nothing positive, inspirational or edifying to say about the United States of America. They believe and are now teaching that America is racist, mean, unjust, immoral, homophobic, imperialistic, xenophobic and just about any other bad adjective you can come up with. If that's true, why are throngs of people from around the world still trying to come here? How many are likewise pouring into China, Russia, Iran, Saudi Arabia, Nigeria, Venezuela, Pakistan or India?

Additionally, students are taught in some schools that America is the source of most of the world's problems, and our military has been imperialistic. The truth is that the American military has likely been the most benevolent in history, with the goal of liberation and freedom, not colonization, rule, and domination. Most leftist academics and college students have no idea of the importance, power, and deterrent force of the formidable U. S. Navy in the world today, not to mention our other military branches. College students are taught untruth in many disciplines today; much of it is preposterous, and often the fruit of evolutionary materialism, political correctness, radical feminism, and identity politics. It was a shame not too long ago that former Secretary of State Condoleezza Rice felt she simply had to withdraw from Rutgers University as commencement speaker, with all the liberal outcry and protests. Ms. Rice is an amazing woman, an African-American of achievement and accomplishment, but was just *unacceptable* to the sophomoric undergraduates because she's borderline conservative and served in a Republican administration. It just gets awkward on those campuses when *the inmates are running the asylum*.

We know from history and the Bible (Deuteronomy 12:8, and Judges 17:6; 21:25) that it never works out when everyone does what is right "in their own eyes," leading to disunity, immorality, strife and chaos. Just look around you, in today's nearly post-Christian America. What do you see? Don't be surprised; after all, culture does seem to be religion (or the lack thereof) externalized.

Many of our greatest early statesmen and leaders attended these formerly Christian schools, and were instrumental in the founding of perhaps the greatest country in the history of the modern world, regardless of what the academic leftists and social justice warriors tell you. So what happened? How and when did the puzzling about-face get started? There are many contributing factors, but if you had to give a one-word answer, it actually is very simple: *evolution.* I mention elsewhere in this book that among the protagonists, pundits and scientists involved in the origins debate, it is widely agreed that there are *only two* real options, or choices, of how life forms, including human beings, came to be. Either living organisms somehow made themselves from non-living materials and chemicals, or they (and us) were made by an omnipotent uncaused First Cause—a Creator God. There really is no other option. We either have a divine lawgiver, with moral absolutes and profound long-term ramifications, or we evolved from the primeval slime, are just another type of animal, and everything is relative—with no basis whatsoever—for any universal moral guidelines. H. G. Wells, the well-known science fiction writer of the early 20th century, was also a social and political commentator. While accepting the claims of Darwinian evolution, and an avowed atheist, he uttered a succinct commentary on the issue a century ago: "If *evolution* is true, *nothing* matters. If the *Bible* and Christianity are true, *nothing else* matters."

An astute statement indeed by Mr. Wells, who at times seemed so insightful, and other times not, like his sexual liaison with atheist racist Margaret Sanger (as countless other men had), the infamous founder of Planned Parenthood (more on that organization later). He also was one of the pundits who liked to describe WWI, after its conclusion in 1918, as *the war to end all wars* (racist and progressive President Woodrow Wilson made the same claim). That bold statement proved highly inaccurate just two decades later with the cataclysm of WW II; always be skeptical of the "experts." This critical subject of origins will be explored much more as we look at real facts, real history and real science. As the conversation continues on the suppression of American men and authentic manhood, and the new call for an end of supposed "toxic masculinity," along with the assault on traditional marriage, this rise of materialism and evolution must be considered.

It will become apparent that the unjustified, lightning-fast adoption of Darwinian evolution, by educational elites and secular pundits in the late 19th century onward undergirds all of these problems, and indeed our current regrettable and baffling culture war. Most of you young men and women out there want to succeed and do the right thing, but I strongly suspect widespread confusion while awash in this secular sea of political correctness and relativity. Do you know what is right? Do you live on bed-rock unchanging precepts, or have you fallen for the lie that morality is a personal choice? Do you know how a real man or woman should act? In considering these all-important questions, let's begin with a volatile issue that has recently burst upon the American scene, one that is loaded with misinformation, manhood assault, and censored history.

3

CONFEDERATE MONUMENTS, CENSORED HISTORY, AND REAL AMERICAN MANHOOD

AS MANY STRUGGLE for air in our confused culture covered with a blanket of political correctness, the suddenly volatile issue of long-present monuments and statues honoring Confederate heroes and Generals has burst upon the scene. It is not surprising that the most well-known and respected of our 18th-century Founding Fathers have been thrown into the firestorm as well. Most of these statues have been up for many decades, some for a century or more, and they have inspired generations of Americans—south and north, black, brown and white—and especially young men. Hopefully we can still consider the wisdom of Englishman Samuel Johnson: "a contempt of the monuments and the wisdom of the past, may be justly reckoned one of the reigning follies of these days, to which pride and idleness have equally contributed." It is only *recently* that the political left's identity politics-victimhood-white supremacy-apology culture and mantra has reached a loud crescendo on this particular issue, thus this sudden, nothing more than feigned and copycat outrage is a leftist fake, and savvy citizens should recognize this. Misinformation and historical ignorance abounds. Even African-American born-again Christian conservative commentator and author Star Parker, who on other issues deserves credit and recognition. She recommended after 2017's volatile summer monument controversies that President Trump should explain the difference between George Washington and Thomas Jefferson, contrasted with the likes of Robert E. Lee and Stonewall Jackson, to the American people-as if there really is one- and she knows and understands what it is. I appreciate Ms. Parker, especially her work related to urban renewal and the abortion issue, but strongly suspect she knows *next to nothing* about these two Confederate generals, *much less their views and actions on slavery*.

Following a tumultuous, even violent weekend of demonstrations in beautiful and historic Charlottesville, Va., back in 2017 related to removing a statue of Robert E. Lee, noodle-backboned government bureaucrats and officials in more than ten cities quickly removed various Confederate monuments and markers. As usual this was against the people's will, as subsequent polls showed a majority wanted the monuments to stay. Robert E. Lee is honored and memorialized with a stained glass window in the Washington Cathedral, and with a statue in the U. S. Capitol. We have more than 10 U. S. Army bases named after Confederate generals, including Ft. Jackson, Ft. Hood, Ft. Gordon, and Ft. Bragg, not to mention other military entities, including warships. The Robert E. Lee ballistic missile submarine was decommissioned after 31 years of service in 1995. We also have the intimidating guided missile cruiser *Chancellorsville* (named after one of Lee and Jackson's great victories); there are others. Should we scrap all of these as well? The official painting of the military's chaplain's corps is a beauty of General Stonewall Jackson with pastor Beverly Tucker Lacy. Has it got to go, too? Do you often wonder how such a small minority of the *hate America* crowd wields so much power? It will be explained in this book. And while we're at it, how about the attack on sports teams, like Washington "Redskins," Cleveland "Indians," Atlanta "Braves" and a multitude of other similarly-named athletic teams? On another tangential issue, many social justice "warriors" today would be surprised that New York native and Republican President Theodore Roosevelt (early 1900s) was not only the first president to invite a black American to the White House-the great Booker T. Washington-but the eclectic northern president often had one of his favorite tunes, "Dixie," played at formal state dinners, and considered it the most popular of our national anthems and songs. People loved it-nobody ever complained. Where does the ridiculous political correctness and divisiveness stop? I've come across at least four national commentators or speakers in the last two years that used the term "cold civil war" to describe our frustrating cultural impasse; let's hope it never morphs into another *hot* one-that would be catastrophic.

When I was a boy growing up in West Virginia, we had access to really cool, well-done Civil War trading cards for several years, similar to baseball cards. My brother and I and a few buddies treasured these with their great depictions of various soldiers and generals-north and south-in action, with colorful battle scenes, including a descriptive historical anecdote on the back of each card. We were so disappointed when they were discontinued. They were informative and inspirational for young boys, and we each had our favorite dudes, north and south. Boys and young men (and young women) need inspirational heroes-or models-to duplicate and pattern their life after. We would reenact as best we could, and take turns being Yankees or Rebels. Civil War stuff was a part of our life. My precious mother (now in her 80's) was even head cheerleader at *Stonewall Jackson* High School. I personally have been to many Civil War battlefields on multiple occasions, especially the sacred ground of Virginia, and several of the key combatants (north and south) were inspirational models for me in my move toward manhood.

To honestly and objectively dig in and discuss this Confederate statues issue, we need to look at real historical facts and especially the men who fought in this epic struggle, and how they felt about not only the war and slavery, but each other. Even the Allies' inspirational leader to victory in WWII, Winston Churchill, whose mother was American, said that he never read about or witnessed such courage by both sides in a major conflict, as that manifested in the American Civil War. He was particularly impressed with the titanic week-long struggle at the "Bloody Angle" of Spotsylvania Courthouse. The American Civil War is unique in world history. There were some 620,000 casualties in the bloody conflict-more American casualties than in all other American wars combined-sometimes with brother fighting against brother, friend against friend. Even Abraham Lincoln's wife had brothers that fought for the Confederacy. The intrepid cavalry leader of Lee's vaunted citizen-army, "Jeb" Stewart, was married to the daughter of one of the Union's top cavalry commanders. Important Union General John Gibbon, who saw intense action in many battles throughout the war, had three brothers and a well-known second cousin (scholar, linguist, and lawyer General J. Johnston Pettigrew) fight for the Confederacy.

On numerous occasions, best friends before the war faced off, like Generals Lewis Armistead and Winfield Hancock on Cemetery Ridge at Gettysburg, and Ulysses S. Grant opposing James Longstreet in the last year of the war. The latter was best man at Grant's wedding *before* the war. I have a beautiful and poignant print above a fireplace in my home by Civil War artist Dale Gallon, called *Born To Be Brothers*. It shows various Confederate and Union officers mingling and amicably talking in the wet yard of the McLean house at Appomattox in April, 1865, right after Lee's surrender to Grant inside the McLean house (Lee is not present; he had left a short time before). The central focus of the painting is Union Commander Grant standing outside, reaching out to touch Lee's then second-in-command James Longstreet's elbow, while offering his old friend a cigar, and saying, "Let's have another game of *Brag* to recall the old days" (Brag was a simplified form of poker). They had been shooting at each other just a few days before. Longstreet was understandably moved, and later after much reflection commented, "Why do men fight who were born to be brothers?"

The respect, friendship, and honor represented in this scene occurred repeatedly between combatants at war's end and thereafter, but most movingly at the formal surrender of Lee's remnant army there at Appomattox a few days later. The surrender of Lee's remaining ragtag battle-hardened troops, to the impressively-arrayed formal ranks of thousands of federal soldiers, was planned and accepted by intrepid Union General Joshua Chamberlain. Chamberlain was one of the heroes at Gettysburg nearly two years before, as his 20th Maine regiment anchored the critical end of the Union left flank on Cemetery Ridge. American manhood was in full display by both sides on that hot July day in 1863, at a now legendary rocky hill called Little Round Top. He was a strong Christian, married with two young children, a college professor proficient in 9 languages and with no military training before the war. He was the only Union soldier awarded a battlefield promotion by General Grant during the war (June, 1864), this due to valor leading a desperate charge (ordered by Grant) on *Rives' Salient* at Petersburg, which Chamberlain obediently carried out but knew was doomed to failure. He was severely wounded and expected to die, but amazingly recovered enough to eventually return to duty. Relative to the surrender of Lee's army, President Lincoln instructed Grant to "let 'em up easy", and Grant agreed but nevertheless felt that there should be some sort of organized formal ceremony with a stacking of arms. General Chamberlain (later 4-time Governor and 2-time Senator of Maine) gave deep thought and prayer throughout the night as to how he would handle the surrendering Confederate troops, who he had great respect for, as he imagined those tattered but worthy regiments marching up the hill to officially surrender, stack arms and part with their treasured battle flags. After much prayer and a sleepless night-thinking about the right thing to do-and even lightly considering possible repercussions, he made his mind up.

The leading Confederate regiment of the near-starving southern army quietly and loosely marched up the hill that damp April morning, still proudly carrying their tattered but beloved battle flags. They were led by the indomitable but now downcast General John Brown Gordon (later Governor and Senator from Georgia) on his fine steed, trailed by the decimated but legendary Stonewall brigade, which started the war with some 4,500 enthusiastic western Virginia young men, but was then reduced to 200. As they reached the stopping point, there suddenly occurred a startling crisp Union bugle call, followed by an immediate measured shift and precise sound by the smartly formed lines of thousands of federal troops. Gordon and his following men immediately perked up from their lethargy. One soldier described it later as like "a shock of electricity" running through the troops. General Chamberlain had surprisingly ordered an honorary *soldier's salute*—arms first to the shoulder—then presented forward—an amazing tribute of respect which the surprised General Gordon immediately recognized.

The southern General, tall and lanky but normally ramrod-straight with an inspirational powerful voice, who one soldier described as "the pertiest thing in a saddle you ever did see," reared his dark stallion up on its back legs in response, drew his sword, and touched the toe his boot with the tip of his sword . . . as an honorary return salute, as his following troops now held heads higher, backs a bit straighter, and rifles raised smartly to the shoulders. Chamberlain had his well-formed troops give the same salute to *every* Confederate regiment, and General Gordon ordered it returned by the opposite Confederates, regiment after regiment, honor for honor, all in otherwise total silence. As one who has studied Gordon and Chamberlain extensively (who were, amazingly, the two main orators in the country for reconciliation after the war), and been to the actual site where the surrender took place numerous times, I get a lump in my throat and a few chill bumps every time I read about it, or visit the site. What mutual respect-what powerful character-*what American manhood* . . . no bands playing, no cannons or fireworks fired off by the victorious Union soldiers, no yelling, no taunting, no trash-talking, no finger pointing. This profound event probably helped save the country.

General Joshua Chamberlain had confided to just one of his officers concerning his plan; when he gave the go-ahead and the bugler called for the unique salute, even the impressively arrayed Union troops were likely surprised. Chamberlain thought he might get court-martialed, but he later and always stated: "It was the right thing to do." It most likely did cost him at least the Vice-Presidential nomination in 1868 and 1872 of the Republican Party, but claimed if given a second chance, he would do the same thing again. It was truly an incredible moment not only in American history, but in the annals of warfare, as well as representative of the mutual respect of the combatants for each other. On display was American manhood at its finest-no snowflakes present, no gender-confused, no wimpy retreat to college *safe places*, no verbal put-downs. Even gritty Union Commander General Ulysses S. Grant, when later writing with respect about those scarecrow southern soldiers, described them as "that enemy, whose manhood, however mistaken the cause, drew forth such herculean deeds of valor."

It is well-known that many comrades-in-arms that honed their skills and fought side-by-side in the Mexican War (1846-1848) opposed each other in the American Civil War. This War Between the States had indeed a unique wartime dynamic, and it wasn't that long ago. The last living documented soldier of that sanguinary conflict didn't die until the 1950's. I was surprised back in the early 1990's while scanning the Charlotte Observer newspaper at the announcement of the recent death of Stonewall Jackson's granddaughter. The American Civil War and Reconstruction were singular events of worldwide importance, but they are now just glossed over in most public-school American history books and classes, and worse yet, have become victims of revisionist history.

Slavery is one of the key issues, but misinformation abounds. Exact figures are unknown, but my research and that of others indicates that at least 85% of Confederate soldiers never owned a slave. Most were hardscrabble or yeoman farmers or small-town young men that had never been more than thirty or forty miles from home. From letters sent home and post-war comments from those that marched north across the Mason-Dixon line into the scenic (to this day) southern Pennsylvania farmland in the Gettysburg campaign, it is obvious they were mesmerized, if not stunned, by that northern land of plenty and especially the large, well-built and beautiful stone barns scattered across the landscape. These lean and confident, sometimes cocky southern boys, probably one third of 'em barefoot by this time (end of June, 1863, 90+ degree heat), would probably have laughed at you (or worse) if you'd accused them of fighting to defend slavery. As one soldier's letter illustrates, during a lull in the fighting, while carefully eyeing his young counterpart while exchanging northern coffee for southern tobacco in the shallows a Virginia river, Billy Yank asked Johnny Reb, "Why are you fightin'?!" His answer: "Cause *you're here!*" Remember, the North invaded the South. Yes, southern states had seceded, but this was not a new concept; attempts and applications for secession had first been introduced several times by disgruntled northern states the previous 50 years, going all the way back initially to disagreement on the War of 1812.

Today's advocates of Confederate monument and statue destruction and elimination are offended at honored recognition of southern officers, who supposedly were all nasty traitorous racists fighting to defend and uphold slavery. Is that picture historically accurate and a proper perspective, though? Let's look at a few of the key Confederate generals that are among the recent targets of monument madness, their character, and importantly their views and actions on slavery, particularly two of the most targeted and famous warriors. Readers would likely be hard-pressed to find the following information in revisionist American history books, or hear it from liberal college professors, with their ever-present politically-correct *thought police* roaming the hallowed halls of academia.

After much thought and prayer, Robert E. Lee turned down an offer to command the *federal* army at the outbreak of hostilities, claiming he could never turn his sword against his family and home state of Virginia. Thus he was not only turning down the prestigious command of the Union Army; he was essentially turning down the U. S. Presidency as well. Back then Americans made victorious generals their president. Many historians feel the war would have lasted at most only 6 months if Lee had accepted the Federal command; the South had no trained army. Lee's family had been Virginians for 290 years, Americans for just 80. It is difficult for us moderns to understand this state-focused loyalty and mindset, especially considering our experience with the current omnipresent behemoth of an all-powerful national government. But even Thomas Jefferson, well into the 1820's and to the end of his life, referred to his beloved Virginia as "my country." Lee's home was what is now known as Arlington, where the famous national cemetery resides on the edge of Washington, D. C. It was confiscated by the federals at the beginning of the Civil war, an action ruled illegal by the United States Supreme Court in 1883, thirteen years after Lee's death.

He inherited this property through marriage to Martha Washington's great-granddaughter Mary Anna. He was surrounded by many paintings, furniture, artifacts, and memorabilia of George Washington's in this home at Arlington. Washington was his model, or hero, growing up; he never saw his own Revolutionary War cavalry hero father, "Light-horse Harry," after age six. Harry had numerous weaknesses and challenges, including investment and land speculation problems. The *Father of our Country*, however, was fond of him and Harry did write, at Washington's request, the famous eulogy for Washington's own funeral. "First in War, First in Peace, First in the Hearts of his Countrymen" later became etched into American history, but Lee's father could not deliver it himself as he was in debtor's prison, eight years before Robert E. was even born.

The earlier Lee colonial family was praised as among the most prominent and important in America's early history, even by George Washington, and two of them signed the Declaration of Independence. As a sidebar, it was another great Virginian (who died the same year as Washington, in 1799), Patrick Henry, who earlier warned that the new Constitution as written proposed (or at least allowed) too strong a central government, that would likely promote regional strife and differences, and quite possibly lead *to civil war*. George Mason, who had argued for an end to the slave trade, felt likewise, and partially due to their efforts the Bill of Rights was added (these two statesmen were sharp-even prophetic).

Lee graduated number one overall in his class, and second in academic ranking-by a fraction of a point-at the United States Military Academy at West Point, and to this day is the only cadet to ever make it through the storied institution without a single demerit for citizenship, rules or behavior infractions. He was one of the most outstanding engineers to ever serve in the well-known Army Corps of Engineers, and is in large part responsible for construction and improvements at Ft. Monroe, Ft. Pulaski, and saving the city of St. Louis from the encroaching Mississippi River (a truly amazing feat brought in $20,000 under budget and several months early). He eventually served as the Superintendent of the United States Military Academy at West Point for several years in the early 1850's, and then Colonel of the Cavalry in dangerous Texas in the late '50s. Lee spent over 50% of his military life away from his wife and family in dedicated service to the United States government and military, but always without even a hint of sexual impropriety, even though he very much enjoyed the company of women and was called more than once "the most handsome man in America." (Incidentally, the famous beard didn't appear until the Civil War). Commanding General Winfield Scott,

Lee's commander in the Mexican War (1846-1848), said Lee (then a captain) was perhaps the finest soldier in the field he had ever served with, and probably the most influential in the American victory in Mexico. Later General Scott stated that if given the chance, Robert E. Lee would probably be the greatest military commander yet in American history.

Lee believed slavery was a moral and social evil, and was a proponent of general manumission of the slaves. As there had been eight generations of slaves, he felt *immediate* emancipation would only multiply that evil. Lee had inherited nearly 200 slaves through marriage, describing it as "a distasteful thing." Lee felt the gentle caresses and influences of Christianity need be applied, as they first needed education, preparation and guidance. Lee also felt strongly that the institution of slavery had within it the negative notion that work is undesirable (so get other people to do it for you). He actually thought it produced even greater deleterious effects on the slave *owners*. Virginia had the largest number of freed blacks among southern states just prior to the war-more than 60,000- and was against secession, as was Lee, but Lincoln's stunning call-up in early 1861 for 75,000 troops to invade the South was simply too much for the proud Virginians. As Lee felt the Negro people needed to be educated and prepared first-to avoid social chaos-he didn't just talk the talk-*he took action*. With the aid of his wife and daughters, he established a *home school* in Arlington for the slaves he inherited through marriage, hoping that by doing it this way, in their home, they wouldn't get thrown in jail for violating education restrictions. Following the Nat Turner slave rebellion in 1831, when 51 whites were ruthlessly slain in two days, the new stringent oppressive legislation prohibited the public education of black bondsmen.

General Lee believed very strongly that the slaves needed three things: 1)-an education, 2)-a marketable skill or job, and 3)-ownership of property. By 1857, three and a half years *before* the outbreak of the Civil War, Lee had manumitted (prepared and freed) over 160 slaves, and signed papers freeing the last one in 1862 (he had inherited 196 through marriage). Following home education, with his daughters doing most of the teaching, he helped them all develop a skill and get located, usually with property. Conversely, although magnanimous to Lee's army at the Appomattox surrender, Grant did not free his own few slaves until 1867, two years *after* the war ended. Grant's #2 man and federal commander in the more western and southern theatres of operation, William T. Sherman (loathed by some southerners to this day), initially condoned slavery and sympathized with the south, but he interestingly chose to side with the Union when war erupted "because it was the clearest symbol of the order he craved," according to biographer John Marszalek. One wonders if this information about Lee is found in any of the revisionist public school history books used today, or mentioned by politically correct National Park Service guides on Arlington tours. I wouldn't bet on it.

General Lee and other army officers from the South showed little enthusiasm for a military solution to the stunning secession crisis, and were usually less vocal and more reluctant to leave the Union than the populace. Lee's solemn own words:

> I take great pride in my country, her prosperity and institutions, and would defendany state if her rights were invaded. But I can anticipate no greater calamity for the country than the dissolution of the Union . . . Whatever may be the result of the contest, I foresee that the country will have to pass through a terrible ordeal, a necessary expiation perhaps forour national sins.[19]

New President Abraham Lincoln was inaugurated on Mar. 4th, 1861; just one month later, on Apr. 4th, the important state of Virginia voted 89 to 45 *against* secession. Following the bombardment of Ft. Sumter, S.C. eight days later by hot-headed Confederates, Lincoln quickly issued a call-up a few days later for 75,000 troops to invade the South and put down the rebellion. To say that the two most important states in the South, North Carolina and Virginia—who were **against** secession- -were ticked off by the brand new rookie President's hasty action would be a huge understate-ment. Virginia did a prompt about-face and voted to secede two days later, followed by North Carolina, Arkansas and Tennessee in May. It seems the classic example of things quickly spiraling out of control; whether Lincoln too aggressively jumped the gun is an interesting topic for debate.

Many folks thought the war, when and if it came, would be over in weeks or a few months, maybe even after one conclusive battle. Hundreds of young fellows were worried it would be over before they got into action, and they would miss out on all the glory. Lee, who well knew the might of the United States military and northern industrial power, is the only one on record who prophetically predicted at least 4 years of bloody struggle.

Lee's Christian character and humble spirit were recognized by all who knew him. The following is revealing in a letter to his wife the second year of hostilities: "I can only say that I am nothing but a poor sinner, trusting in Christ alone for salvation. I tremble for my country when I hear of the confidence expressed in me. I know too well my weakness, and that our

---

[19] J. William Jones, *Life and Letters of Robert Edward Lee Soldier and Man*, Sprinkle Publications, Harrisonburg, VA, 1978, p. 13 (reprint, 1909)

only hope is in God." It is difficult to imagine a national leader or politician speaking like this today. After the war and declining many lucrative job offers, even from large northern companies and European governments, Lee accepted an offer to direct little Washington College (now Washington and Lee University) and invest his energies into young men. His simplification of the honor code there consisted of just one rule: *A student must behave as a [Christian] gentleman*. His first campus improvement was to build a beautiful chapel. General Lee, who had taught or served with many of the northern officers, could never bring himself to refer to the opposing northern army as "the enemy," and stated in a post-war conversation that there was never a day he didn't pray for them.

One simple example of Lee's heart was illustrated well into the war when several hungry officers enthusiastically presented with three chickens and a half dozen potatoes-gifts from a local farmer- anticipating a rare delicious meal. General Lee instructed them politely to keep a couple of potatoes for soup, and take the rest down to the suffering men at the hospital-possibly searing the officers' conscience. Perhaps one can sense Lee's spirit and noble heart by this well-known eyewitness account (from A. L. Long and Marcus J. Wright, *Memoirs of Gettysburg*):

> I was at the battle of Gettysburg myself, and an incident occurred there that largely changed my views of the Southern people. I had been a most bitter anti-Southman, and fought and cursed the Confederates desperately. The last day of the fight I was badly wounded. A ball shattered my left leg. I lay on the ground not far from Cemetery Ridge, and as General Lee ordered his retreat, he and his officers rode near me. As they came along I recognized him, and, though faint from exposure and loss of blood, I raised up my hands, looked Lee in the face, and shouted as loud as I could, "Hurrah for the Union!" The general heard me, looked, stopped his horse, dismounted, and came toward me. I confess that at first I thought he meant to kill me. But as he came up he looked down at me with such a sad expression upon his face that all fear left me, and I wondered what he was about. He extended his hand to me, and grasping mine firmly and looking right into my eyes, said, "My son, I hope you will soon be well." If I live a thousand years I shall never forget the expression on General Lee's face. There he was, defeated, retiring from a field that had cost him and his cause almost their last hope, and yet he stopped to say words like those to a wounded soldier of the opposition who had

just taunted him as he passed by! As soon as the general had
left me I cried myself to sleep there upon the bloody ground!

Robert E. Lee often applied Scripture; perhaps in that poignant moment he was moved by Proverbs 24:17 and 25:21. You young and millennial men and women across the Heartland, understand that it's all about character. In J. Steven Wilkin's *Leaders in Action* <u>Call of Duty</u> book on Lee, a wonderful read, he has as his sub-title *The Sterling Nobility of Robert E. Lee*. I cannot imagine a better sub-title, or a more appropriate phrase, to describe this great man. Another example of Lee's character was exhibited in a Sunday worship service at St. Paul's Church in Richmond barely two months after the war ended. Over half of the burned city was in ruins, while uncertainty, distrust and tension permeated the atmosphere, especially since the recent assassination of President Abraham Lincoln. The city was under martial law. That early summer Sunday morning a free black man, ushered in and egged on by two Union soldiers, stunningly went forth *first* for communion to the front of the sanctuary in the packed church. In the Anglican (American Episcopal) Church they would do communion at the end of every service; whites would normally go first, then the blacks (upstairs) would go second, while singing for each other. Seating was actually *not* segregated in southern churches until after the horrific Nat Turner slave rebellion. Union authorities and occupation soldiers were closely monitoring what was said in the churches, and some would do things to provoke Southerners.

One can just imagine the awkward, potentially volatile situation as a hush fell over the packed church. It was Custis Lee (the General's oldest son), when writing about it later, who said that it seemed as if all the air had been taken out of the sanctuary. With all eyes focused forward and the tension reaching an almost unbearable level, the hushed but charged atmosphere was suddenly interrupted by the sound of a measured military cadence-the sound of General Lee's footsteps-as he went forward, and then knelt and joined the black man side by side for communion. The diaries and letters indicate that Lee put his arm around the black man, and then raised his other arm for communion. The rest of the congregation then joined them, black and white together. What a scene it must have been. What could have been a very bad situation was made good by Lee's character. Take note: circumstances do not determine your character; they expose it, and present opportunities to strengthen and refine it. As my friend, pastor and historian Harry Reeder further puts it, "the man doesn't make the moment, and the moment doesn't make the man . . . the moment tells us what kind of a man we are."

Even in defeat, Lee was the most important figure all throughout the South in the first five years after the war (a lofty position he didn't seek), as it recovered from destroyed families, occupation forces and dissension. Lee was strongly against proposals of fleeing westward to the mountain country and carrying on partisan or guerrilla warfare, but instead was all about reconciliation and submitting to God's will. Gamaliel Bradford, in the introduction of *Recollections and Letters of Robert E. Lee* (a book by Lee's youngest son, Capt. Robert E. Lee), described it this way (in 1904):

> His countrymen in all the states of the Confederacy looked up to him as their guide and leader, and few men have ever held a more extensive, unquestioned influence than his . . . every bit of this influence was used for good, for the largest, noblest objects, to restore peace and harmony and quiet, to teach his people not only to submit physically, but to look to the future with the broadest, most genuine, most substantial hope. Where his finger touched, where his word guided, there was to be no bitterness, no lingering grudge, no rebellious hatred. There was just one task for South and North both, and that was to build up a new America that should be wiser, better, and therefore stronger than the old. When a lady, who cherished something of the old hatred, as ladies will, came to him and poured out her complaint, he said to her, in words as noble as were ever uttered by a vanquished hero: "Madam, don't bring up your sons to detest the United States Government. Recollect that we form one country now. Abandon all these local animosities and make your sons Americans."

Sterling nobility, indeed. On another occasion after the war while riding with his daughter Mildred outside of Lexington, home of Washington College (now Washington and Lee University), they stopped by a country home and asked for a drink from the well. Parts of this area had been ravaged by Union General David Hunter. The lady of the small farm recognized General Lee and gave permission, and pointed out a particular dead tree nearby as they stood near the well. She then told the story of how Hunter's men had pulled her husband-a member of the Stonewall brigade- who was recuperating from a battle wound, out of the house and lashed him to that tree and killed him. The bitter lady then shared that every time she looked at that tree, she was again reminded of that terrible incident, the hate boiled up, and she would not cut the tree down. "I hate them!" According to Mildred, as they prepared to leave, General Lee-after some quiet reflection-and in a measured response, said this: "Madam, I am grateful for your husband's service and sacrifice and to you for your kindness. But cut down the tree. Cut down the tree." Lee simply refused to be bitter.

In the excellent book *Lee, The Last Years*, by Charles Bracelen Flood, the author recognizes and thanks in the dedication a particular Virginia gentleman "whose conversations led this northerner (Flood) to realize that this was a story that belonged not only to the South but to our nation as a whole." Absolutely still true; our young men today are in desperate need of such real heroes. Even Martin Luther King Jr. came to admire and respect General Lee, as did the earlier Booker T. Washington and George Washington Carver. In his last few years following the Confederate surrender, even after defeat, Lee was the most famous man in the country, but it was a position he did not desire or really care for. Humility is one of the most important traits of real Christian manhood; see more R. E. Lee examples of this in the later "Toxic men" (Ch. 8) section.

Perhaps some of these monument protestors would feel differently if they would just read and study the issues and real history, including what Lee did for the slaves he inherited through marriage. Unfortunately, they don't know what they don't know. Modern-day African-American writer and educator Edward C. Smith, a strong supporter and admirer of Robert E. Lee, says that "the twin maladies of ignorance and misinformation are not incurable diseases," but then adds, though, that you've "got to read widely and deeply." Oh, we can only hope.

Another Confederate hero targeted by monument radicals, second-rate history professors, and intimidated college administrators, is Thomas Jonathan "Stonewall" Jackson. He was probably the most famous man in the world at the time of his death a week after his unforgettable flank attack at Chancellorsville. There Lee and Jackson, with their much smaller force, just might have bagged Hooker's entire Union army if given just 90 more minutes of daylight. Two months later, at a peaceful crossroads Pennsylvania town called Gettysburg, his absence was sorely felt by Lee's army. Europeans and various other world leaders closely followed the American Civil War, and in the first half of the conflict no star shone brighter than that of Jackson's. His incredible Shenandoah Valley Campaign early in the war, when he and his *foot cavalry* covered 600 miles in 60 days and defeated four federal armies, each larger than their own, is still studied in war colleges not only in the United States but around the world. He was recognized for bravery as a young artillery officer in the Mexican war, especially at a bloody place called Chapultepec, and afterwards taught at the Virginia Military Institute in sleepy Lexington, Va. Jackson became a Christian during his wartime experience in Mexico, and before long, with the guidance and teaching of especially two pastors, he was *all in*. It is virtually impossible to properly understand this unique and somewhat eccentric American hero apart from his Christian faith.

Jackson also felt that slavery had to go and that it was against Christian principles, and he favored manumission, with education and preparation. He felt like he could sidestep education restrictions and the government crackdown by setting up a school for the blacks of Lexington, Va. in his church on Sunday afternoons and calling it his *Sunday school*, thereby hopefully avoiding jail time. He was challenged for this by a man who was a fellow church member, who was not in favor of what Jackson was doing, and claimed Jackson was breaking the law by setting up a school. Major Jackson stood his ground, knowing the state had no jurisdiction over the church, as he informed the man it was just a *Sunday* school. He focused on Bible study, catechism, reading, writing, and arithmetic. One of these black students and his eventual family, educated and helped by Jackson beginning in 1857 (not "General" or "Stonewall" just yet), much later formed their own church near Roanoke, Va. with a unique special window showing a grove of trees and the words "Let us cross over the river and rest in the shade of the trees." These were Jackson's dying words in early May, 1863, a week after Chancellorsville. The special window honors their founder, teacher and mentor- Thomas J. "Stonewall" Jackson- that exists to this day. Jackson had not only helped and taught the black pastor and founder of this church when he was a youth, but helped his parents start a little business. (Did you readers ever learn about this in your politically-correct, revisionist high school or college American history classes?) Those misinformed people that badmouth and categorize Lee and Jackson as racist bigots supportive of slavery are poor students of real history. Incidentally, the gentleman and fellow church member that challenged Jackson on his new "Sunday" school was won over, and ran the school for some twenty years after Jackson's death. A man of integrity stood his ground, determined to help people in need, and good things happened.

Stonewall Jackson was a hero of mine as a boy, even more so today. His exploits in the Shenandoah Valley campaign, 1st and 2nd Manassas, Harper's Ferry, Antietam, Fredericksburg, and Chancellorsville are truly impressive and studied in war colleges around the world. His flanking maneuvers at 2nd Manassas and particularly at Chancellorsville are especially notable. Starting just before dawn on May 2nd, 1863 in the second-growth woods with his entire Corps (some 25,000 men), his axe-men and engineers had to cut much of their own road through brush and woods. Marching 16-17 miles quietly through a mostly narrow trail to get into position on a warm day, his line of troops ten miles long, Jackson was unable to issue his famous soft-spoken attack order to General Robert Rhodes, "You may go forward, then" until 5:15 P.M. His Corps' stunningly successful surprise attack to slam the exposed right flank of "Fighting Joe" Hooker's glorious, well-fed and well-equipped 100,000- man Union army was unparalleled in American military history, later copied by "the Desert Fox" Erwin Rommel in WW2, and then again in the 1990's by "Stormin' Norman" Schwarzkoph in the Gulf War of the Middle East. The Union troops were routed, only darkness prevented annihilation. Historian-Pastor friend Dr. Harry Reeder, with a twinkle in his eye, once said when we were visiting the Chancellorsville battlefield that God knew the South wasn't going to lose the war *with* Jackson-so he simply had to go! Jackson biographer James Robertson talked much of the general's Christian faith, and how Jackson thought the Old Testament and the Book of Maccabees (in the Apocrypha) were the greatest military manuals ever written. He summed Jackson up this way: "When you encounter a man who is fighting with Old Testament fury in order to attain New Testament faith, the only way you can defeat him is to kill him."[20] Which is what happened" (by "friendly" fire, in the dark woods at Chancellorsville). That perspective reminded me of the statement by Walter Taylor, one of Lee's top aides during the war, from his book *General Lee* (p. 125). Taylor was referring to the all-important and infamous lost copy of Lee's battle order at Antietam, which allowed Union Commander McClellan to know how and where Lee had weakened and divided his Army (but amazingly could still only manage a draw), but I think in a larger sense applies to the entire conflict:

> To me it is as if He who controls the destinies of men and nations had said: "You, people of the South, shall be sorely tried, but the blame is not yours, and therefore to you shall fall the honors-genius, skill, courage, fortitude, endurance, readiness for self-sacrifice, prowess in battle, and victory against great odds; but this great experiment to demonstrate man's capacity for self-government, with its cornerstone of universal freedom, must continue with undivided front, and therefore I decree to the other side [the North] determination, persistence, numbers, unlimited resources, and ultimate success.

Misinformation abounds about slavery, the Civil War, and some of the main actors. Possibly a few honors, medals, namesakes and statues may be undeserved, as follows every war, but many of these monument men

---

[20] For the best comprehensive work on Stonewall Jackson, consider *Stonewall Jackson: the Man, The Soldier, The Legend, by James L. Robertson, Jr., Cumberland House, Nashville TN, 1997*

were flat-out American heroes, and inspiration for millions of young men. **Leave the monuments alone**.

This new cultural indicator of Confederate monument assault and removal merits no more time and space in this small book. In closing I would be remiss, though, if not briefly contrasting the Christian character and war perspectives of Lee and Jackson with a handful of their prominent northern counterparts. Ulysses S. Grant, William T. Sherman, John Pope, Phillip Sheridan, David Hunter, and George Armstrong Custer-just to name a few-endorsed and implemented a heretofore unpracticed and un-American *Total War* policy and strategy in the Civil War. This included the **abuse and destruction of non-combatants: women, children and old folks**, along with their property. There were **thousands** of such casualties. This was stunning to Lee, Jackson and the Confederate leadership, just like the news of Union General Joseph Hooker's policy of bringing in prostitutes (thus, the word *hookers*) to his huge army's winter camp, in the months before his massive Army of the Potomac got embarrassed by Lee and Jackson at Chancellorsville. Incidentally and regarding real manhood and character, in the fine book *Chancellorsville 1863*, author and former Marine Corps Colonel Ernest Ferguson makes the convincing point that never in the history of American warfare has *character* so determined the outcome of a major battle. The cocksure womanizer "fighting Joe" Hooker was intimidated by the audacity, character and courage of Robert E. Lee; to his credit, he even admitted as much later.

Returning to the *total war* policy of the Union leadership, the Confederate officers strongly felt that war was just for combatants-or warriors-and warriors only. General Lee's armies had three invasions during the war into northern territory, in 1862 (Antietam), 1863 (Gettysburg campaign), and 1864 (smaller force led by Jubal Early near Washington, D.C.), and there was only one documented civilian casualty-*only one*- when a stray ricocheted bullet came through an open kitchen window (or doorway) and killed Ms. Jenny Wade in Gettysburg. Lee's philosophy, found written in one of his notebooks: "The duties exacted of us by civilization and Christianity are no less obligatory in the country of our enemy than in our own." What a contrasting worldview and policy to that of Generals Grant and Sherman. Particularly in Sherman's infamous "March to the Sea" from Atlanta to Savannah beginning on Nov. 15th and thru Dec. 21st, 1864, he led 60,000 Union soldiers nearly 300 miles in his *scorched earth* program. Property destruction was in the many millions, with virtually all installations, buildings, homes, and infrastructure destroyed with several thousand civilian casualties (black and white), along with rape and pillage, as Sherman lost command and control. Stark naked chimneys (*Sherman's matchsticks*) can occasionally still be seen in the Georgia countryside to this day.

The sanction of this horrific *Total War* philosophy and strategy by Grant, Sherman and other northern officers rocked the Confederate leadership, particularly Jefferson Davis, Lee, John Brown Gordon, and—prior to his May 1863 death—Stonewall Jackson. Among Sherman's instructions to his men, just before that horrific human juggernaut, was *"Make them howl!"* Grant's order to Philip Sheridan in late summer 1864, with the war's end almost in sight, was to turn the beautiful Shenandoah Valley into *a barren waste*. When Sheridan finished this awful destruction of farmland, barns, and many homes, with the capture all the livestock he could find, and the abuse of non-combatant citizens, the little general bragged that "a crow would have to pack three days' rations to make it across that valley!" John Pope and David Hunter's similar *total war* policy earlier in Virginia and elsewhere shocked General Lee, and the closest example ever found of the southern commander using profane language was when he described General Pope as a "miscreant" (Lee and Jackson would not use or allow profanity). Pope was the egotistical general who bragged, after given command of the Army of the Potomac in the summer of 1862, of what he was "going to do to ol' "Bobby" Lee," as Pope's men were used to seeing "only the backs of the enemy" out west. As the self-aggrandizing Pope finally collided with Jackson, Lee, and Longstreet for the first time a month later on August 28-29, 1862, near Manassas Junction (2ⁿᵈ Battle of Bull Run), the only thing that kept Lee and Jackson from destroying and capturing Pope's entire Union Army, of over 50,000 men, was a heavy rainstorm and darkness; it was an utter rout.

Was this heretofore un-American, *total war* policy justified, and successful (as opposed to the concept of Christian *"just"* war)? Many would claim otherwise. From these federal generals just mentioned came the men most responsible for perhaps the darkest shadow, the greatest blight, in American military history-**the abuse and slaughter of Native Americans** (the western Indians)-from mid-1865 to the mid-1880's, with the same appalling policy. Just check your history, and look at what Sherman, Sheridan, Hunter, Pope and the self-centered George Armstrong Custer did out west—after the War Between the States. It is stunning, tragic, and shameful . . . and a permanent large stain on the history of the United States Army.

George Washington, Thomas Jefferson, and other Founders' statues and monuments are also now under attack; it is beyond ridiculous. People need to take a stand against these radical and socialist students, and particularly their **radical professors and administrators** that are leading the attack on America's very Founding and heroes. They are destroying this country, and it has become painfully obvious that they are out to further rewrite American history. Those earlier mentioned Confederate generals were great men, great Americans, and an inspiration for millions of boys, girls and young men. Let their monuments and statues stand. Let us move on and discuss those amazing Founders, and the important issue of slavery.

4

THE FOUNDING FATHERS, SLAVERY AND KING GEORGE

THE FACT THAT some of our Founding Fathers had slaves has now become to many on the left and all social justice warriors an immediate, unequivocal and non-negotiable invalidation of anything courageous or significant they may have accomplished, written (even the Declaration and Constitution), risked, or even heroically performed on the battlefield. It is regrettable, unjustified and agenda-driven, with misinformation and historical ignorance everywhere. The late eighteenth century American era was world-changing, not only because of the sublime but audacious Declaration of Independence, along with the American Revolution, Constitution, and the unique Republic it birthed, but this Founders era was the initial turning point in American slavery.

It should first be stressed that slavery had been introduced nearly from the get-go in the early colonies, going back nearly 150 years before the Founder's era. It was *not* introduced by the Founders themselves.

For you readers who are victims of revisionist history, the *Founding Fathers* were those men of character who were leaders and shakers in the Revolutionary War for Independence and the American Founding, and the subsequent establishment of these United States as a new self-governing nation, which has become arguably the greatest source and force for good and hope in the world the last 200 years, perhaps of all time. Looking at real history—sometimes hard to find in college classrooms today—and the rise and fall of kingdoms and nation-states through the ages, it is stunning what the still-young United States of America has accomplished and produced in this short time period while spreading the message of hope and liberty around the world. Some of the so-called *Founders* signed the key founding documents, like the fifty-six signers of the Declaration of Independence; others did not but were absolutely key actors in this singular world-changing historical plot. This includes the incomparable *Son of Thunder*— Patrick Henry- -the long-time Governor of Virginia who was the *trumpet* of the Revolution. Also included in the Founding Fathers are the fourteen stout-backboned short-term patriot Presidents. This was an impressive group most modern-day students were never taught about who led fledgling America and the Continental Congress from 1774 to 1789. This was mostly during the War for Independence and the short Articles of Confederation period; more on a few of them shortly. Without going into a detailed breakdown of the generally accepted Founder's group, **it is generally agreed that there were 240-250 Founding Fathers**, and despite leftist revisionist history claims by secular academic elites, nearly all of these men openly and unashamedly described the new United States of America as a Christian nation. Time and space does not allow a comprehensive detailed proving of such a claim in this book; it has been done already by numerous honest and respected historians. One quote though, from a patriot even George Washington considered as valuable as any other in the success of the American Revolution-Patrick Henry:

> It cannot be emphasized too strongly or too often that this great nation was founded, not by religionists, but by Christians; not on religions, but on the gospel of Jesus Christ. For this very reason peoples of other faiths have been afforded asylum, prosperity, and freedom of worship here.[21]

For the reading-challenged who might consider the purchase of just one such book relative to this subject, *Original Intent* by David Barton (WallBuilder Press) is highly recommended.

The unjustified secular attack and outcry, especially by left-leaning authors, pundits, and academic elites along with their myrmidons—the strident, unhappy, godless undergraduate students— against our Founding Fathers and Documents involves two major areas of focus: 1) slave ownership, and 2) morality. This includes the false contention that most of them were marginal Christian believers or even atheists—deists at best—and the claim of no influence or preference for Christian principles in government. Additionally, the Founders are often portrayed by these revisionist historians as cruel, greedy dudes and even sometimes immoral womanizers. Nothing could be further from the truth.

Surprisingly to many, this aggressive attack on the Founding Fathers, even our founding documents-indeed, the very *Founding* of the United States of America- started over a century ago by liberal academics and pundits, after the lightning-fast enthusiastic acceptance of the new secular

---

[21] Bradford, M.E. *The Trumpet Voice of Freedom: Patrick Henry of Virginia.* Marlborough, N.H.: The Plymouth Rock Foundation, 1991.

religion, Darwinian evolution, particularly in colleges and universities. This started in the 1860s and by the turn of the century, well before WWI, the foundational transformation to naturalism (with the atheism that always follows) was virtually complete in elitist circles, although not among the less-enlightened commoners and hard-workin' citizens across the fruited plain (many of your distant relatives and mine).

One well-known early example of harsh criticism of Christianity and proponent of revisionist history of our Founding was by Robert Ingersoll, a skilled lawyer and popular speaker who strongly advocated agnosticism, free thought and humanism in the late 1800s. Although many Christian believers then considered the popular Ingersoll quite the pagan in post-Civil War America, he nevertheless attracted quite a following due to his undeniable oratorical skills, as he spoke on a wide variety of subjects, but his favorite seemed to be agnosticism and criticism of Christianity. This critic was one of the most well-known individuals and influential secular culture warriors in the latter part of 19th century America, but not one many people today know much about. He strongly agreed with one of his long-deceased heroes, Thomas Paine, and especially his contention that mere *un*inspired men—not God—had written the Holy Bible (antithetical to *theopneustos*, or "God-breathed," which refutes any claim of human inspiration). Ingersoll attained celebrity status as an eclectic speaker, social pundit, and political commentator. He seemed to despise Christianity— especially apparent after the late 1870s— with the catalyst of Darwinian evolution spreading like a wildfire through academia and seemingly backing his views. Ingersoll's hatred of the church was also partly due to a negative church experience he observed his father go through when just a lad. His popular speeches and presentations along with his secular worldview even influenced industry leaders, and the up and coming so-called *robber barons*, as social Darwinism was embraced ("only the strong survive" and "survival of the fittest") and led to the abuse of thousands of workers. With his popularity and oratorical ability, his false historical claims unquestionably influenced receptive college professors, societal elites and commentators as the new evolutionary paradigm swept through academia. One example:

> [O]ur forefathers retired God from politics. . . The Declaration of Independence announces the sublime truth that all power comes from the people. This was a denial, and the first denial of a nation, of the infamous dogma that God confers the right

upon one man to govern others . . . our fathers founded the
first secular government that was ever founded in this world.[22]

This is simply stunning to me, especially considering the influence
Ingersoll apparently had, as the Declaration of Independence-our most
important founding document-mentions God or Creator *four* times. Historian
David Barton cites another very similar false historical claim[23] by liberal historians Charles Beard and his co-author and early feminist wife Mary; I mention
it here because the Beards were so influential with their revisionist historical views and writings, which helped lay the foundation for today's social
and political chaos. One false claim from their 1930 book, *Rise of American
Civilization*: "national government was secular from top to bottom," and that
our Founding Fathers had "rear[ed] a national government on a secular basis."

The Beard husband-wife team, who were also strongly influenced
by Darwinian evolution, would sometimes resort to outright falsehoods
(including the one just mentioned) to promote their secular claims and
attacks on the Founders. An example: "In dealing with Tripoli [relative to
Muslim piracy, terrorism and trade], President Washington allowed it to
be squarely stated that "the government of the United States is not in
any sense founded upon the Christian religion."[24] This is false because
Washington never made such a statement, there is no record of such a
statement, and he never actually saw the Treaty of Tripoli, and it did not
even arrive in the States until months after he left office. Incidentally, is
it not ironic that we are still dealing with much more serious Muslim terrorism today, 230 years later? Fortunately in those days, President Thomas
Jefferson finally put the hammer down on those Muslim pirates.

Confirming what was previously mentioned, historian David Barton
emphasizes that based on our founding Declaration document alone the
United States Supreme Court noted that our government was *not secular*
since in numerous references it invoked God and His principles into civil
government.[25] I could go on extensively with examples of now disproven
revisionist history claims, their collective agenda to discredit and take down

[22] Robert Ingersoll, *Ingersollia, Gems of Thought*, edited by Elmo (Chicago: Belford,
Clarke & Co., 1882), pp. 49-54.

[23] David Barton, *Original Intent*, WallBuilder Press, 2000, p. 279.

[24] Charles A. Beard and Mary R. Beard, *The Rise of American Civilization* (New York:
The MacMillan Company, 1930), p. 439.

[25] Ibid, the Beards, *The Rise*, p. 280.

our Founders, and thus invalidation of our Founding Documents, as well as their denial of the powerful influence of Christianity on America's Founders. A major point: such revisionists have been around for quite a while, but with a big momentum boost in recent years from unconstitutional judicial activism, political correctness, support from the secular mainstream media, and now additional, often surreptitious support through Big Tech and biased social media manipulation.

Let us look briefly at some key Founders and their views on Christianity and slavery, starting with Henry Laurens of South Carolina, who was a very active and important Founder and one of fourteen "Presidents" of the fledgling United States (men who directed the Continental Congress between 1776 and 1789) before George Washington became the first President and the new Constitution was finally passed. Laurens was one of the most influential Founders most modern-day Americans have never heard of, following the better-known John Hancock as President of the struggling new United States in 1777 during the Revolutionary War, and then ambassador to the Netherlands in 1779. As the new ambassador Laurens was unfortunately captured by the English on the high seas, and imprisoned in the infamous Tower of London for the duration of the war, our only president ever held as a prisoner of war by another country. As a measure of his importance and influence, he finally was released by the Brits in a prisoner exchange for the British military commander Lord Cornwallis himself, and Laurens was once described in that tumultuous period by even George Washington as *the father of our country*.

Laurens, although from agricultural South Carolina, was no fan of slavery. This is from his days in the Continental Congress:

> I abhor slavery. I was born in a country where slavery had been established by British Kings and Parliaments as well as by the laws of the country ages before my existence. . . . . In former days there was no combatting the prejudices of men supported by interest; the day, I hope, is approaching when, from principles of gratitude as well as justice, every man will strive to be foremost in showing his readiness to comply with the Golden Rule ["do unto others as you would have them do unto you" Matthew 7:12].[26]

---

[26] Frank Moore, Materials for History Printed From Original Manuscripts, the Correspondence of Henry Laurens of South Carolina (New York: Zenger Club, 1861), p. 20, to John Laurens on Aug. 14, 1776.

Key Founder John Adams despised the institution of slavery:

> [My] opinion against it [slavery] has always been known . . .
> [Never] in my life did I own a slave.[27]

Numerous articles and books have unjustly attacked and denied the importance and influence of biblical belief, and the Christian character of, the more well-known Founding Fathers. Virtually all of these writers, critics and pseudo-historians were agenda-driven atheists (like the Beards). Again, a powerful statement from our 2nd President, John Adams:

> The general principles on which the fathers achieved
> independence were . . . the general principles of Christianity
> . . . I will avow that I then believed, and now believe, that
> those general principles of Christianity are as eternal and
> immutable as the existence and attributes of God; and that
> those principles of liberty are as unalterable as human nature.[28]

Returning to the slavery issue, the perception of it as a more complicated issue in the South in the late 1700s due to agriculture and economic concerns is probably warranted. Even the Virginia Assembly, though, in 1772, seriously and formally addressed the English King in protest *against* the slave trade. Patrick Henry, the inspirational leader of leaders in the Virginia colony and a kind and courageous man, who unquestionably was one the most important Founding Fathers (and a hero of mine), seemed often distraught over the issue of slavery. From a follow-up letter to a Quaker friend, after reading a book on the slave trade, part of his response follows:

> It is not a little surprising that the professors of Christianity whose
> chief excellence consists in softening the human heart, and in
> cherishing and improving its finer feelings, should encourage

---

[27] John Adams, *The Works of John Adams, Second President of the United States*, Charles Francis Adams editor, Little, Brown and Company, Boston, MA, 1854, Vol. IX, pp. 92-93.

[28] Ibid, Adams, *The Works*, Vol. X, pp. 45-46

a practice so totally repugnant to the first impressions of right and wrong? What adds to the wonder is that this abominable practice had been introduced in the most enlightened ages. . . Is it not amazing, that at a time, when the rights of humanity are defined and understood with precision, in a country, above all others, fond of liberty, that in such an age and in such a country, we find men professing a religion the most humane, mild, gentle and generous, adopting a principle as repugnant to humanity, as it is inconsistent with the Bible, and destructive to liberty? Every thinking, honest man rejects it in speculation, how few in practice from conscientious motives![29]

Henry biographer David Vaughn, in his wonderful book *Give Me Liberty*, points out that Henry continued on, rebuking his own practice of holding slaves: "Would anyone believe I am the master of slaves of my own purchase!. . . .I will not, I cannot justify it," and he hoped for an opportunity "to abolish this lamentable evil,"[30] while preparing and waiting for the acceptable and appropriate time for emancipation, while doing his best to "improve it." Indeed, slavery was part of life, the economy, and the way of the world in the western hemisphere (as well as elsewhere). It almost seems to us moderns as if the southern colonies got to a point where they were trapped between a rock and a hard place (along with the forced policies of King George) with the expansion and growth of agriculture and plantations. A friend and I used to go in by small boat in the winter, not too far from Savannah, Georgia, and bow-hunt for wild hogs in the unique and wild coastal salt marsh habitat of that area. I was amazed when he would point out the hard to see, but still existing woodworks and jetties constructed centuries ago for the original rice plantations (almost 300 years old).

When cotton and tobacco replaced rice as the primary cash crops, slave labor was often viewed as even more of an economic necessity. This was especially so as the 19th century quickly dawned for the new nation and cotton became *king* in the southeast, with receptive and hungry European markets. Even with this complicated issue, we need to *follow the money*—a rarely discussed topic (which always puzzled me). Attempts and petitions by some colonists and Founders to curtail or eventually stop slavery always were blocked by King George and the British leadership. It is interesting that

---

[29] David J. Vaughn, *Give Me Liberty*, George Grant general editor, Highland Books, Elkton, MD, 1997, p. 278.

[30] Ibid, pp. 278-279

even Thomas Jefferson, while also a slaveholder, included a denunciation of slavery in his original draft of the *Declaration of Independence*, but it was removed as a concession to those delegates from the aforementioned southern states where the labor of black bondsmen "seemed essential to the economy." Remember also at this time, war was right on the threshold for these rebellious colonists, despite having *no real army or navy*, against one of the most powerful nations in the world. Britannica still ruled the seas in many areas, and its army had not lost a major land battle in ages; they were unquestionably a force to be reckoned with. Most of these American leaders and Founding Fathers were distracted, to put it mildly; those who signed the Declaration of Independence in July of 1776—an act of treason— had, for all intents and purposes, just signed their death warrant.

Senior sage and Founder (and always seemingly loveable) Benjamin Franklin frustratingly lamented the deaf ears of the King and Parliament of England to American appeals and petitions to end slavery, in this 1773 letter:

> . . . a disposition to abolish slavery prevails in North America, that many of Pennsylvanians have set their slaves at liberty, and that even the Virginia Assembly have petitioned the King for permission to make a law for preventing the importation of more into that colony. This request, however, will probably not be granted as their former laws of that kind *have always been repealed*.[31]

The multi-talented patriot Thomas Jefferson has also come under unjustified attack (and his monuments) by revisionists and social justice radicals. One of the most memorable and edifying commentaries on Jefferson and his dislike of slavery came from northerner John Quincy Adams, who was quite the talented and eclectic scholar, statesman, patriot, and President himself:

> The inconsistency of the institution of domestic slavery with the principles of the Declaration of Independence was seen and *lamented by all the southern patriots* of the Revolution; by no one with deeper and more unalterable conviction than by the author of the Declaration himself [Jefferson]. No charge of insincerity or hypocrisy can be fairly laid to their charge. Never

---

[31] Benjamin Franklin, *The Works of Benjamin Franklin*, Jared Sparks editor, Tappan, Whittemore, and Mason, Boston, MA, 1839, Vol. III, p. 42.

from their lips was heard one syllable of attempt to justify the institution of slavery. They universally considered it as a reproach fastened upon them by the unnatural step-mother country [Great Britain] and they saw that before the principles of the Declaration of Independence, slavery, in common with every other mode of oppression, was destined sooner or later to be banished from the earth. Such was the undoubting conviction of Jefferson to his dying day. In the *Memoirs of His Life*, written at the age of 77, he gave to his countrymen the solemn and emphatic warning that the day was not distant when they must hear and adopt the general emancipation of their slaves.[32]

Real American history shows that in the earliest days, from the early 1600's up to the 1760's, prior to the Revolutionary War period (as British colonies), attempts to end slavery were few and far between. Important Founding Father, President of the Continental Congress, Federalist Papers author, Bible Society member and 1st Chief Justice of the United States Supreme Court John Jay admitted as much, while then intimating the time and reason for the genesis of a changing American mindset toward slavery:

Prior to the great Revolution, the great majority . . . of our people had been so long accustomed to the practice and convenience of having slaves that very few among them even doubted the propriety and rectitude of it.[33]

That all changed with the spirit of freedom and calls for liberty with the eventual revolt against King George. The American Revolution was the cata-lyst for the change in the colonial view about slavery—the "turning point"— writer David Barton calls it, and he discusses in detail how the Founding Fathers contributed greatly to such a signal attitude adjustment, as they increasingly and vigorously complained, like Ben Franklin mentioned pre-viously, "against the fact that Great Britain had forcefully imposed upon

---

[32] John Quincy Adams, An Oration Delivered Before The Inhabitants Of The Town Of Newburyport at Their Request on the Sixty-first Anniversary of the Declaration of Independence, July 4, 1837; Charles Whipple, Newburyport, 1837, p. 50.

[33] John Jay, *The Correspondence and Public Papers of John Jay*, Henry P. Johnston, editor, G. P. Putnam's Sons, New York, 1891, Vol. III, p. 342.

the Colonies the evil of slavery,"[34] here almost brutally described by the soft-spoken Thomas Jefferson's elegant but scathing pen:

> He [King George] has waged cruel war against human nature itself, violating its most sacred rights of life and liberty in the persons of a distant people who never offended him, captivating and carrying them into slavery in another hemisphere or to incur miserable death in their transportation thither. . . . Determined to keep open a market where men should be bought and sold, he has prostituted his negative for suppressing every legislative attempt to prohibit or to restrain this execrable commerce [that is, he has opposed efforts to prohibit the slave trade].[35]

It is indeed a complicated issue, but I do tend to agree with the words of author John Perry: "To see the truth in history, we have to resist the temptation to hold past actions accountable to contemporary standards."[36] This rejuvenated attack on America's Founding Fathers, and even the *very Founding* of this great nation is unjustified, distorted and ignores real history, very similar to the Civil War heroes and monuments controversy. Perhaps the modern secular radicals' attack on our early great statesmen and leaders is related to the Founder's prevailing view on *who* should direct the new nation. Here is a well-known example from brilliant Federalist Papers author and first Supreme Court Chief Justice John Jay: "Americans should select and prefer *Christians* as their rulers." Many American Founders who had been slave owners as British colonial citizens released them after the split, and as previously touched on, many never even owned slaves. Was there procrastination, disagreement and delay on the resolution of the slavery issue, along with growing pains of the infant new nation as it moved into and through the early 19th century? Absolutely; it is a complicated issue and multiple parties are at fault, which sadly contributed to the convulsive sanguinary reckoning of the American Civil War.

People need to stand up and let their voices be heard. Others need to first simply ***wake up***—and realize that there are those out there who are

---

[34] Ibid, Barton, *Original*, p. 289

[35] Ibid, Barton, *Original*, p. 290, quoted from Thomas Jefferson, *The Writings of Thomas Jefferson*, Albert Ellery Bergh, editor, Thomas Jefferson Memorial Association, Washington, D.C., 1903, Vol. I, p. 34.

[36] John Perry, *Unshakable Faith-Booker T. Washington and George Washington Carver*, Multnomah Publishers, Inc., Sisters, OR, 1999, p. 11

out to destroy this nation—led by secular academics and media elites. Even now as I finish this commentary, the liberal bureaucrats of San Francisco are wrestling with a decision. The city with many thousands of street people, downtown tent cities, human trash, spent drug paraphernalia and human excrement blanketing the sidewalks and parking lots everywhere are up in arms over what should be a non-issue. Their bureaucrats are trying to figure out which option to pursue to eliminate or cover up a now *controversial* mural of George Washington—at George Washington High School—after *a few* activists complained. The options proposed include simply painting over it for $600,000, masking or covering it with paneling for $875,000, or concealing it with large curtains for a measly $300,000. The claim of "traumatized" students (from viewing the mural) is a joke, as the school's students (according to National Review) "were *against* its removal or just apathetic." Although in my high school days I was probably more interested in sports, riding my dirt bike, and girls, I actually had a feeling of pride that I was able to attend George Washington High School (in Charleston, W. V.). I mean, that dude was the Father of our Country, and even did the original surveying of the Kanawha valley in western Virginia when a young man, where the city of Charleston is. What school could possibly have a better namesake?! Unknown to many Americans and rarely openly discussed, another part of the monuments/Founders attack by the left, though, is George Washington's (and the other Founder's) views and actions on *sodomy*—now termed homosexuality. The first time that a sodomite (homosexual) was kicked out of the American military was by Commander-in-Chief George Washington in the American Revolutionary War:

At a General Court Martial whereof Colo. Tupper was President (10th March 1778), Lieut. Enslin of Colo. Malcom's Regiment [was] tried for attempting to commit sodomy, with John Monhort a soldier . . . [he was] found guilty of the charges exhibited against him, being breaches of 5th. Article 18th. Section of the Articles of War and [we] do sentence him to be dismiss'd [from] the service *with infamy* [*public disgrace*]. His Excellency the Commander in Chief approves the sentence and with abhorrence and detestation of such infamous crimes orders Lieut. Enslin to be drummed out of camp tomorrow morning by all the drummers and fifers in the Army never to return.[37]

---

[37] Washington, Writings, (1934), Vol. XI, pp. 83-84, from General Orders at Valley Forge on March 14, 1778

The eclectic and internationally-known, honored and respected Thomas Jefferson, another key Founder, proposed *castration* as the penalty for sodomy (homosexual behavior).[38] This is the same Thomas Jefferson who more than once mentioned that one of his main worries, concerning the new American Republic, was the possibility of the Courts overstepping their authority, and instead of *interpreting* law would begin *making* law—a judicial oligarchy—an unconstitutional rule of a few over the many. What impressive prescience and understanding of power; he would be stunned at our *judicial activism* problem today! Important and influential Founder James Wilson, who emigrated from Scotland when a young man, but who eventually signed the Declaration of Independence and the Constitution and became one of America's most important statesmen and initial Supreme Court justices, was so repulsed by sodomy (again, that was the term used two centuries ago), that relative to his legal profession he stated:

The crime not to be named [sodomy], I pass in total silence.[39]

How things have changed. With the casting aside of biblical authority and the reality of our sexually permissive culture, now including not only legalization of homosexual marriage but promotion of gender fluidity and transgender rights, is it any wonder that the attack on our Founding Fathers and even the founding documents has accelerated? The very term, "Founding *Fathers*," in our gender-confused and sexual culture, and any use or insinuation of male and/or female, is supposedly very offensive to transgender people—the "*Fathers*" part particularly. Parents need to think thrice about where they send their kids to college, and even *where and how* they educate their children pre-college. These radicals hate America, they dislike and have no respect for our founding documents, many don't respect the rule of law, they don't believe in a transcendent Creator God or moral absolutes, and many mock and despise Christianity. Brilliant lawyer, Congressman, and orator Daniel Webster often paced in his office before important Supreme Court cases and presentations while reciting Scripture from the ancient Book of Job. He was a strong Christian and was formerly described as the *Defender of the Constitution*. In a

---

[38] Thomas Jefferson, *Notes on the State of Virginia*, Matthew Carey, Philadelphia, PA, Query XIV, 1794, p. 211

[39] James Wilson, *The Works of James Wilson*, Belknap Press of Harvard University Press, Cambridge, MA, 1967, Vol. II, p. 656

powerful July 4th presentation in the state of Maine in 1802, and still a young man, Webster said this about the relationship between morality and religion:

> To preserve the government we must also preserve a correct and energetic tone of morals. After all that can be said, the truth is that liberty consists more in the habits of the people than anything else. When the public mind becomes vitiated and depraved, every attempt to preserve it is vain. Laws are then a nullity, and Constitutions waste paper. . . Ambitious men must be restrained by the public morality: when they rise up to do evil, they must find themselves standing alone. Morality rests on religion. If you destroy the foundation, the superstructure must fall.

The superstructure hasn't fallen yet, like it has in Europe, but there are major cracks in it. Two decades ago an older and wiser personal friend and patriot, with an impressive military and business background, told me—with great foresight—that the only way America can be defeated *is from within*. Is it now happening before our very eyes?

The adoption of ungrounded, unscientific, never observed (in the past, present, or laboratory) macro- or Darwinian evolution—which goes against the most proved laws of science—with the casting aside of biblical mores is a major part of the problem and at the foundation of this radical, strident, sometimes belligerent atheistic movement. And don't think for a minute that these groups (just like some of the Central and South American *immigrant* groups) don't have financial backing from leftists with deep pockets. You might consider this harsh-sounding language, but like another one of my childhood heroes, the *Say Hey* kid—the great Willie Mays—used to say, "no brag, just fact." Like many Americans, I love this country (and its monuments)—warts and all—and what it stands for. I want folks to know some truth. They're hard-pressed to get it from the mainstream media, Hollywood, public school textbooks and higher academia or entrenched government bureaucrats. And of course, no worries at this end; I can't get fired.

5

MANHOOD, MARRIAGE, TIME AND GENETICS

MANY MILLENNIALS AND young people today doubt the historical reality of a supernatural creation by God of the first pair of humans, Adam and Eve. Most do not know that human genome and mitochondrial DNA research essentially prove that all people trace back to one pair of human beings-one man and one woman.[40] It was amazing when this information came out several years back, but it was generally ignored by the mainstream media and higher education (including some evolutionary biologists), just like the recent multiple findings of still-soft and flexible, unfossilized dinosaur tissue. This screams young age-*thousands* of years, not millions or billions-and is also supported by radiocarbon dating. Research chemist, Christian writer and speaker Bruce Malone (Search For the Truth Ministries) refers to Carbon dating as an especially big "elephant in the living room" for evolution believers because things we bring up out of the sedimentary rock layers, made from formerly living life forms— like coal, diamonds, petroleum and incompletely hardened fossils—*all* have significant and measureable levels of radioactive Carbon 14, which has a short "half-life." Anything older than a measly 150,000 years or so should *not have any* Carbon 14 left in it! Students have been brainwashed for a century by academics and evolutionists that the dinosaurs died out 65-70 million years ago. Multiple un-fossilized dinosaur finds in the last dozen

---

[40] Nathaniel Jeanson, PhD, *Replacing Darwin-Made Simple*, Answers in Genesis, Hebron, KY, 2019, pp. 34-38; also Julie Von Velt & Bruce Malone, *Inspired Evidence*, Search for the Truth Publications, 2011, April 26th reading; Nathaniel T. Jeanson, *Replacing Darwin*-The New Origin of Species, Master Books, Green Forest, AR, 2017, pp. 207-280; and Nathaniel T. Jeanson, PhD, *Does "Y-Chromosome Adam" Refute Genesis?*—from *Acts and Facts* Magazine, Institute for Creation Research, vol. 42, #11, Nov., 2013, p. 20. Also, for the genetic esoterica-inclined, Dr. J. C. Sanford's *Genetic Entropy* (FMS Publications, 2014) is highly recommended relative to this subject, and will lead the reader to additional sources and references.

years or so (including stunning finds of even intact *Dino DNA fragments*) should have been front-page news in every major news outlet in the world. These findings are powerful evidence for *young* age of the earth, dinosaurs, fossil fuels, diamonds, human beings and everything else. Every time these new dinosaur finds are tested, they contain measurable levels of unstable carbon 14. They couldn't be millions of years old. This evidence is seemingly irrefutable; many evolutionists simply *ignore* it.

There is no conceivable explanation how such sensitive organic molecules, from dead animals, could still be extant in the rock layers for the many million to over two-hundred million years timescale postulated by deep time evolutionists. The only rational explanation is rapid catastrophic mud-water burial on a massive scale just thousands of years ago, where the buried organisms are shielded, so to speak, from weathering, oxidation, burrowing animals, etc. Some 75% of the earth's land is composed of sedimentary water-deposited rock layers that look like they were of large-scale flood origin. So what gives, with the evolutionists? The answer: **Nothing, It doesn't fit their worldview** (so *don't bore me with the facts*).

Not surprisingly, secular scientists involved in this particular fascinating biochemical information and genetic research the last 20-25 years do not refer to this founding couple as Adam and Eve, but such researchers who are Christians do believe that. Regardless of a geneticist's religious leanings, is it not profound that such research into the tiny and heretofore hidden control centers of human cells, where the chromosomes and genes-our DNA-reside tucked away in obscurity, with an information software code that dwarfs anything designed by man, has revealed that all human beings have come from *one set* of parents? The Bible is not a science book per se, but it often touches on science, and every time it does, it is amazingly prophetic and accurate (see in **Part IV**, *When the Bible touches on Science*). I suppose we shouldn't be too surprised that it has taken man thousands of years to confirm what the Bible says. As a matter of fact, as skilled researchers finished mapping the human genome and moved on to other animals, they found the *same* thing-genetic evidence of only one original pair, *one set* of parents, male and female, in the animals as well. Some of you may recall, in Genesis 6:19, God commanded Noah, ". . . *two* of every sort shall thou bring into the ark, to keep them alive with thee, they shall be male and female." (Noah didn't have to go find them, God sent them; in the next verse He said "they shall come unto thee"). These animal genome findings were significant and surprising, but the particularly interesting discovery was yet to come. When they got to genetic mapping of certain *domestic* animals, the scientists discovered that the genome and DNA data indicated that these few species (sheep, goats, etc.) likely originated from seemingly *three to five* sets of parent animals! *Only* these . . . what was going on . . . or, why?

Again, let's look in the Bible, specifically Genesis 7:2. God further commands Noah, "Of every *clean* beast thou shall take to thee by sevens, the male and his female: and of beasts that are not clean by two, the male and his female." The "clean" kinds of beasts (and birds-see verse 3) were those for domestication and more varied use (sheep, goats and cattle) as well as for sacrificial offerings. "Three of *only* these few "clean" types (or possibly pairs) were preserved most likely to allow for greater variation in breeding in the depleted and desolate environment after the Flood. The seventh was apparently offered by Noah, as mentioned in Genesis 8:20, as a sacrifice to God after they disembarked from the ark. (*Note: It is not revealed in Genesis 1 how many original pairs of *clean* animals, or any other type (besides man), were originally created by God).

Do you understand the significance here? Amazingly complex, recent genome mapping supports what the Bible says and predicts: Two original parents for all human beings and most animals, except for just a few *clean* (domestic) animals-which go back to three pairs -". . . by sevens, the male and his female"; i.e., three pairs of parents (and one extra for sacrifice)-at least on the Ark. There is definitely room for discussion here, but it seems another fascinating example has been found that what we read in God's Word seems to agree with what we find in God's world.

Returning to marriage . . . where does it actually come from? Marriage originates in Genesis; it is defined and ordained by God. If one believes that, then it follows that God's Word is the only standard for defining proper marriage. Many of you have heard this at some point, but increasing numbers of young adults in post-modern America have not, because of growing up in secular households with no faith-based worldview, along with their exposure to the higher academia assault on traditional marriage and the family. Just look at our culture—homosexual *marriage* is now legal by a 5 to 4 Supreme Court vote. Do politicians, judges, celebrities, homosexual activists and radical feminists have the right to redefine marriage and family, the most important of all human institutions? It would have been unthinkable just a few decades ago. Genesis 2:24 says, "Therefore shall a man leave his father and his mother, and shall cleave unto his wife: and they shall be one flesh." The same thing is restated in the New Testament in Ephesians 5:31. It is also mentioned again by Jesus in Matthew 19, in responding to those irritating Pharisees, where He again repeats the meaning of marriage: one man and one woman. 1st Corinthians 7: 1-13 specifically offers guidance on marriage. You young and millennial men are an important part of my target audience with this book, so I want to make sure you have some knowledge about marriage, its origin and importance—I want you to succeed in marriage—and be strong and courageous, and be great fathers. Tragically, women in some quarters and countries are now encouraged to have children without marriage and with no biological father in the home, or even involved with the children. This usually leads to serious social problems. Other sexual-related problems abound in our permissive society, like state health officials' challenge in curtailing the growing number of sexual diseases—including syphilis—because of the increase in anonymous users of dating apps. But the evidence is strong: children do best when raised with a biological father.

In light of today's heightened sensitivity to the conventional use of gender-indicating pronouns and nouns, it should be pointed out that the word *man* (when often used for mankind, or human beings)— or *anthropos* in the Greek—which is used some 550 times in Scripture—has nothing to do with male gender. Back to Jesus in Matthew 19, responding with perhaps a hint of polite criticism to the *elites* (I do it occasionally myself) in verse 4, he asks, "Have ye not read, that he which made them *at the beginning* made them male and female?" Jesus pointedly stresses that the Godhead made Adam and Eve right at the virtual beginning, not some imagined 5 billion years afterwards. Here Jesus himself quotes as His authority the creation account in Genesis (1:27), and in his next statement to them refers to Genesis 2:24 when he reminds them that "a man shall leave his father and mother, and shall cleave to his wife, and be one flesh," and then, "What therefore God hath joined together, let no man put asunder." So I encourage you, even if you've never cracked a Bible open, get the one off the shelf that's there to ward off evil spirits, blow the dust off, and read these passages. It is so important. *This is where marriage comes from.*

What we're dealing with not only in America but across the western world is a *disrespect and loss of biblical authority.* The all-important question: *Why?*

The answer is discussed throughout this book. It is because of the ungrounded (**no scientific evidence for**), promotion and acceptance of *Darwinian evolution* (or *naturalism*). I also think of it as *white lab coat intimidation.* Students and people throughout the last century or more have been propagandized and sent the message that science and evolution have disproven the Bible, particularly the important foundational historical narrative of Genesis. Nothing could be further from the truth.

People have been indoctrinated that millions and billions of years of bloodshed, death, disease, struggle and evolution really happened, was a good thing, and somehow produced mankind from that mess. There **is no scientific evidence for this idea, in the present world or in the fossil record**. Most of you have been taught there are gobs of evidence, like transitional forms in the fossil record, strong links between apes and man, or evolution-supporting genetic mutations, but this is false: **there are none**. This book and many others profile and document this deception, but you won't be exposed to them in your science textbooks or classes, or even school libraries because of censorship. It is also significant that there is *not a hint* of long ages (millions or billions of years) anywhere in the Bible. I stress this because of biblical compromise and promotion of the so-called *gap theory*, the *day-age theory*, *theistic evolution* and others by some pastors, intimidated seminary professors and other Christians. These ideas started with controversial Scottish Presbyterian academic, political economist and minister Thomas Chalmers in the early 1800s. In large part this was due to the long-ages/uniformitarian ideas and influence of lawyer, amateur geologist and contemporary Scottish pundit Charles Lyell's 3-volume *Principles of Geology*. Incidentally, these were the only books, besides a Bible, taken on board the ship *Beagle* by Charles Darwin in his now famous five-year survey voyage around the world from 1831 to 1836. This famous journey ultimately birthed his controversial but nevertheless world-changing *Origins* magnum opus and its evolution idea. Charles Lyell, who essentially became a mentor and coach to Darwin, promoted the idea of long-ages/uniformitarianism and surreptitiously tried to explain everything without God. He was profoundly influential on the thinking of the young divinity and medical school dropout-turned-naturalist Charles Darwin, who was later mad at God anyway because of the death of his daughter. In Darwin's mind, here was the big question: if slow and gradual processes over many millions of years applied to geology (Lyell's view), then must it not also apply to biology? I will give Darwin some credit; he was a good naturalist.

Hopefully you will come to realize that the creation account is the original source and foundation for the all-important institution of marriage, specifically discussed and emphasized by Jesus himself, and meant to be a lifelong union between one man and one woman. It is well-known that it is not always smooth sailing, even among Christians. Part of the reason can be found in the second part of Genesis 3:16, where after describing the particular manifestation of the Curse on Eve for her rebellion (Adam's comes in the next verse), i.e., pain and sorrow in childbirth, God says "and thy desire shall be to thy husband, and he shall rule over thee." This is interesting, and is one of many verses mocked by the humanists and liberal left (at least the few who've read it), but in going back to the original Hebrew it actually means that the woman is going to want to exert her will, or try to be the leader in the marriage relationship, but henceforth the husband would be the ruler in the family. I strongly suspect this might be the least favorite verse, among thousands, of the radical feminists. As stated earlier, even in good marital relationships, spouses occasionally butt heads.

I find it interesting that over 600 years ago even the 'Father of English Literature,' Geoffrey Chaucer, best known for the *The Canterbury Tales*, said this: "Women have one consuming desire, the desire for command."

Men and women are wired differently; biblically, they have equal worth and value, but different roles. Equality does not mean interchange-ability. It is a scientific and social fact that men and women are different; it feels so unbelievably ridiculous and time-wasting that this even needs to be discussed, or debated with screaming social/gender 'justice' warriors. On many issues, men in general just don't have that unique perspective, or *women's intuition*; it took me a while to grasp this as a young man. The right reality: equal value, equal importance, but different roles. If you think about it, all successful organizations— corporations, smaller businesses, sports teams, military units, or on-track churches—all have some type of hierarchical order with different roles and levels of submission, account-ability, and communication. Have American men sometimes dropped the ball as spiritual leaders of their families in these modern times? Absolutely, irrefutably—let's talk about one of the reasons.

6

MANHOOD MENACE: PORNOGRAPHY

FEW SOCIAL JUSTICE warriors would have any interest or opinion, but most Christians would agree that the book of Romans is one of the most important books in the Bible. The very first chapter presents a serious introduction as it discusses the power of the gospel, and then the tragic descent of ancient man from godly principles into atheism, pantheism and polytheism, with the always accompanying widespread immorality and egregious sexual sin that follows (see Rom. 1:21-28). When the Creator God is cast aside, people start to quickly focus on and worship the creation, or the "creatures" (same word in the Greek-see Rom. 1:25) rather than Him. This is seen all around us today, whether it is omnipresent self-worship, self-esteem programs, self-image books, discs, DVDs, talk shows, the new but already tiring "selfie" culture, environmental wackos and tree huggers. These people contradict themselves at every turn, or seem to be confused. As an example, they may spend a lot of time, significant energy and resources to oppose the construction of a hydroelectric power plant that would greatly benefit a half million people, just to save a tiny remnant population of some little mussel or salamander that 99% of the folks didn't even know existed. At the same time most of these same zealots support the murder of human babies without hesitation via abortion, with more and more of them okay with it even into the third trimester, or *right up to and even after the baby is born.*

I don't recall where I picked it up, but relative to societal changes I remember someone defining *tolerance* as the virtue of the man without convictions, which stuck with me. In this current so-called culture war and after decades of often ungrounded and agenda-driven calls for "tolerance," there has been an explosion of what most Americans, just a few years back, would have considered serious sexual sin. One devastating modern manifestation of and contributor to this is pornography. Hollywood, ACLU types, the far-left, and many mainstream media pundits unsurprisingly disagree, but the tragic truth: marriages and families are being destroyed, men-and to a lesser degree-women, are getting addicted. It has even become a huge problem with teenagers and children, especially in the last decade with easy access to non-filtered computers, smartphones and high-tech gadgetry. Most teens view porn online now, and if they haven't by the time they go off to college they're going to soon see it, even if the dorm rooms have some sort of Wi-Fi filtering. Well over 50% of millennial men view pornography regularly.

I remember reading a fascinating, well-documented editorial in the Washington Times years ago about how the rise and proliferation of pornography was the driving force behind the development and explosive use of the internet (Al Gore's claims notwithstanding). More recent research indicates pornography comprises at least one third to half of all Internet activity, and pornographic websites get more visitors than Amazon, Netflix and Twitter combined. It is not just an American problem; it is a worldwide epidemic. It has even become a problem among pastors and Christians as well; everyone is vulnerable. All humans are fallen and have "feet of clay," and struggle with the same issues others do. Over 20% of youth pastors admit using or struggling with porn, the last figure I saw for senior pastors was a bit higher. Studies show that at least 50% of churchgoing men view pornography regularly. It has even become a serious problem for young women. Barna research indicates that one third of 13-24 year-old females search for and view pornography at least once a month.

It pains me to think of anyone addicted to pornography, but especially the plethora of young men in today's manhood assault environment. Unsurprisingly, this addictive habit makes it more challenging for men to relate to sharp, grounded women. Currently with young men marriage is down, while pornography use is way up. Is there a relationship here? Data indicates there is, as increasing numbers of young men sink into this abyss of sexual sin. It is a barricade to proper family leadership and fatherhood, loving and serving your wife to the full, reaching your maximum potential and productivity with work and career, as well as spiritual growth, inner contentment and joy. Keith Grabill compares it to leprosy:

> When a person allows the entrance of pornography into his or her life, the infection begins. This internal leprosy of pornography is strikingly familiar to the external leprosy that was feared for so long. Leprosy of old had a debilitating toll on the person as it ate away at the outside flesh of the body.

> But the toll of internal leprosy is far worse, as it eats away at
> the inner heart, soul, and mind of a person.[41]

Studies show that when pornographic images and videos are watched, the user's chemical communication system is effected, as powerful neurotransmitters like dopamine are stimulated and released, which *addict* the viewer to what's seen, as these powerful pornographic images transfer from the eyes through the brain in less than half a second, somehow catalyzing an immediate biochemical and micro-structural residual visual-memory pathway. Although further study is needed, Dr. Tim Jennings, a well-known Christian psychiatrist and certified master psycho-pharmacologist, points out that such repetitive behavior creates circuits or pathways in the brain that are there-virtually permanent-and things then fire (like neural impulse image transfer) quicker, or automatically. Thus, they are micro-entrenched, difficult to get rid of, and thus consequentially come addiction and bondage. Author Herman Melville referred to the eyes as the "window to the soul." Guard your eye-gate. It is a huge problem in our society, especially for men, with scantily-clad females at every turn.

Most neuroscientists claim the human brain is still developing in young adults and does not fully mature until the mid-20s. Returning to the previous idea of automatically—firing pathways, and the young adult human brain approaching wiring and software completion (maturation)—while regularly viewing pornography-the ramifications and possibilities for damage, even permanent damage, are profound. We humans appreciate and use all of our senses, but the "eye gate" is most important (especially in men). Mr. Grabill puts it this way: "Pornographic images quickly imprint . . . the brain is structurally changed and memories are created-we literally 'grow new brain' with each visual experience."[42] Regrettably, the evidence indicates that regular viewing of porn actually changes the chemistry and micromorphology, or physical structure, of the brain, and the powerful drug-like effect creates a need for more perverted or extreme pornographic images for continued satisfaction. One can just imagine the potential sexual relationship problems and flawed expectations in such a teenager's life, not to mention sleep deprivation, negatively affected academic achievement or work

[41] Keith Grabill, *Seven Times-Be Free Live Free*, Search For the Truth Publications, Midland, Mi., 2014, p. 7

[42] Ibid, Grabill . . . quoting from, "The Science Behind Pornography Addiction," Judith Reisman, speech, at a Science, Technology, and Space Hearing, Nov. 18, 2004.

performance, and stifled spiritual growth. This can be an oppressive, all-encompassing addiction problem preventing even an approach to a balanced, productive and abundant life. It is difficult to deal with or overcome, in large part due to the chemical and neurological changes in this most important organ in the human body. On another tangent, it should be mentioned that the irrefutable relationship between hard-core pornography and serious crime and negative social parameters has been firmly established,[43] as has its long involvement with *organized* crime. What's to be done?

Besides restrictions, legislation, filters and parental prioritization and oversight, the here-and-now addicted individual faces a big challenge, as does our entire society. Many don't know, and many may deny, but this insidious issue is one of the most destructive in our confused culture right now as families, marriages, relationships, spiritual walks and more are impacted and often destroyed. Young men's march to real manhood is often truncated or stopped in its tracks, as teenage boys may be the biggest users now. Parents need to pay more attention to their kid's hearts and heroes; someone in the culture will, if they don't. Is it really wise to let pop culture, Hollywood, social media and an atheism-based public school system be the teachers of your children? Perhaps it is for you; it depends on *your* worldview and foundational assumptions (see Part III, 'The Problem with Assumptions'). How about more legislation and government intervention? That would help, but it's a complicated issue with First Amendment free speech arguments. The liberal left and Democrat Party are unsurprisingly *not* supportive of changes, limitations or restrictions. The other side is quite different; the Republican Party Platform's formal view of pornography—what they termed "this public menace"—three years ago (2016, shortly before their convention): "A public health crisis that is destroying the life of millions." A handful of states have declared pornography a public health crisis. The truth is many pornography users are addicted. They cannot stop and break this nefarious habit by themselves; this is a heart and soul problem.

It is surprising to many—but true—what author David Powlison claims:

---

[43] Final Report of the Attorney General's Commission on Pornography, Rutledge Hill Press, Nashville, TN, 1986

> **Idolatry** is by far the most frequently discussed problem in the Scriptures. The relevance of massive chunks of Scripture hangs on our understanding of idolatry."[44]

Just to make sure we understand each other, with no communication breakdowns, here is a to-the-point definition of an idol, from *Gospel Treason*:

> An **idol** is anything or anyone that captures our hearts, minds, and affectations more than God.[45]

Pornography is an idol. Everyone battles different idols, virtually every day. I know all of them can be negative (or addicting) whether it is drugs, romance novels, Facebook, gambling or even golf. Pornography has got to be one of the absolute worst, and toughest to shed. I strongly suggest struggling pornography users consider the claims of Jesus Christ, and give Him a chance to help you "be transformed," as your heart and brain need to "be made new," working in discipleship with and alongside Christians. I strongly recommend the short but powerful book *Seven Times-Be Free Live Free*[46], by Keith Grabill, and the DVD teaching curriculum on sexual purity called the Conquer Series[47], hosted by former U. S. Marine fighter pilot Dr. Ted Roberts. Fight the New Drug (FTND), an apparently secular anti-porn organization, might be an option as well, especially for business entities hamstrung and restricted by politically correct guidelines, but many users in bondage need the just mentioned *supernatural* help. With this problem, you can't rush it—it's a process—as your governing core gets reprogrammed and you get rid of this idol of porn. Any readers with this problem, *there is* a way out—it's *not easy*—but you can exit the abyss . . . and have life again . . . and for sure, "more abundantly."

---

[44] David Powlison, "Idols of the Heart and Vanity Fair," *Journal of Biblical Counseling* 13, no. 2, Winter 1995, 1

[45] Brad Bigney, *Gospel Treason*, p. 24, P & R Publishing Co., Phillipsburg, N. J., 2012.

[46] Ibid, Grabill . . .

[47] The *Conquer Series*, at conquerseries.com, 561-681-9990.

7

TRUTH FOCUS

ONE KEY TO real and long-term success not always mentioned involves truth. Some of you sharper millennials and younger adults sense that trying to be honest, doing your best, and being the best you can be is part of success, inner contentment and peace of mind, whether you're a Christian or not. Scripture does say, in Colossians 3:23, "And whatsoever ye do, do it heartily, as to the Lord, and not unto men." That is, not *wimpily*, or *just good enough for government work*, but giving it your best, and for the right reasons. But what are we talking about here . . . and why? It involves honesty and truth— *about everything*. Being truthful *with yourself*: your work ethic, your priorities alignment (proper order: God, family, business or job, then all the rest), with your spouse, your time management, and how you spend your money (self-centered, or non-selfishly), just to name a few. Focusing on truth, honesty, and doing the next best thing also means being truthful about proper passions, your talents, your skills, as well as your deficiencies and weak points. Even great artists and performers often embrace this mandate. I wrote this commentary down from the Sunday newspaper years ago by Philippe Petit, the extraordinary high-wire daredevil who walked the high wire between the pylons of the Sydney, Australia Harbor Bridge, as well as one between the former Twin Towers of New York. He said this: "The true artistic impulse has mothing to do with pleasing the audience—or, for that matter— with pleasing the impresario so you'll get more jobs or more money. That's not art. If you are an artist, you want to create a giant wall around yourself and, inside that wall, *to follow honesty* and your intuition. What the audience will see is a man or woman who is a prisoner of his or her passion, and that is the most inspiring performance in the world."

What is meant by *proper* passions? It goes back to balance and vertical alignment. As an example, despite widespread obesity all around us (42% of the American population, late 2019), too many young men and women are out-of-balance these days with physical fitness, going to the gym, and bein' buff. I loved sports growing up, loved to compete and still exercise, but there is imbalance considering the time, money, energy and effort some people—men and women—invest in fitness, looking good, and wearin' mirrors out . . . part of the modern-day focus on self. Lots of these folks would be better served by exercising their minds more. Another obvious popular outa-balance aberration: many young adults, especially men, have an improper passion with playing videogames. It dominates their life. This was tragically illustrated last year by a disgruntled young man when he lost in a big videogame competition in Georgia, where he pulled out a gun and started shooting, killing two people and wounding others. There are countless other examples of such questionable passions-they're called idols. We all battle them.

Daily and long-term, no matter job type, income level, who your relatives are, how beautiful or handsome you supposedly are or who you know, should not everyone be on a search for right and truth, and regular honest self-evaluation? Constant self-examination, being true and honest with yourself, seems to be a characteristic of great leaders and achievers. It seems Mr. Shakespeare had it right, "To thine own self be true, then thou canst not be false to any man."[48] Particularly for young adults this includes being honest with yourself about your talents and what kind of work you're good at. Mentors from the present, and even indirectly models (or heroes) from history can help and inspire you. Get you some good models from history you can relate to, and try to duplicate them. Most Christians feel that specific God-given talent or ability is an important consideration in figuring out God's will—workwise—for one's life. Of course all work has challenges, but *are you good at it*? Does it, or eventually can it, give you joy, and God glory? It is an important consideration.

Real winners and balanced successful people seem to relish truth, fairness and honesty, and tend to regularly raise the bar on themselves. By *real* winners I am not necessarily referring to well-known sports or film celebrities, nationally-known politicians, wealthy CEO's or hedge fund managers and the like. Some, to be sure, are fine business leaders or performers, great parents and patriots, with a proven track record and *good* fruit on the tree. But questions to be pondered in consideration of real success and character, which always involves right and truth: Who and how many have they helped or influenced positively? Have they sown good seeds? What do their employees, co-workers or relatives think of them? How is their own health and spiritual life? Do they have a solid marriage and well-balanced children, or is it a family with multiple casualties and a trail of tears? I believe we should all be *fruit inspectors*. As mentioned previously it is a timeless principle: a good tree produces good fruit. How about yourself? Admittedly, we are all human and we all make mistakes, and if you're carrying around a burden of guilt there is a way to deal with that. Individuals with legitimate and real well-rounded success are widely respected and inspire others, have balanced lives and good fruit, and live by honesty and truth.

---

[48] William Shakespeare, *Hamlet*, Act 1, scene 3, through Polonius

Life is best when you do what needs to be done, even when not convenient, and always hold yourself accountable. If it's worth doing, or simply has to be done, do it right . . . and do it right the first time. Despite the current cultural confusion, we fortunately still have many of this type across the fruited plain, and I get excited about them—such men and women have courage and character—also known as guts, or backbone— and are good *finishers*. Even if—as my western friends might say—they might be a little salty. But that's okay— a good thing—and a model for wimpy, intimidated and confused young men in our discombobulated politically-correct society. Surprising to many in this confused culture, even on the main lobby wall of the old headquarters building of the CIA is inscribed the following verse (from John 8:32): "And ye shall know the truth and the truth shall make you free." This powerful statement strongly insinuates that there is truth, and where it might be found. Many throughout the ages have wondered if it even exists, and if so where- or exactly *what*- it is (Pontius Pilate, for example). Victorious and inspirational Allied WWII leader and British Prime Minister Winston Churchill felt strongly about the subject, and said this around the beginning of the first world war: "The truth is incontrovertible; malice may attack it, ignorance may deride it, but in the end, there it is." And significantly for believers, God says (3rd John: 4),

I have no greater joy than to hear that my children walk in truth.

Truth and doing the right thing is strong—I think of it as *core-hardening*— and it is quietly and chronically exhilarating. **Right makes might**, as Abraham Lincoln said. This is direct opposition to the awful axiom of Lenin, Stalin, Mao, Hitler and every other godless dictator or socialist government, which is *might makes right* (whoever is in power and has the guns sets the rules). Don't buy into the current mainstream media/academic/entertainment elite's godless mantra that all truth is relative. I believe it was Christian author and pastor John Maxwell who said "Where there is hope for the future, there is power in the present." Thoughts are things. It is a powerful force when you get to the point where you know that you know that you know. As one of Louis L'Amour's western characters said, "there's nothin' stoppin' a man *who knows he's in the right and keeps on acomin'*!

8

BACK TO MEN-ESPECIALLY THE SOMEWHAT TOXIC, SUCCESSFUL KIND

ONE MAXIM FROM former Secretary of Defense and bluntly loquacious military warrior James "Mad Dog" Mattis: "Be polite, be professional, and be prepared to kill anyone you meet." Shortly thereafter, following the interview question, *Who keeps you awake at night*?" . . . his immediate answer: "'Nobody, I keep other people awake." Controversial and headstrong but victorious American WWII General George Patton, the antithesis of sensitive snowflake effeminate man, once famously said "There's no darn glory in dying for your country—the idea is to make the other poor, dumb bastard die for his country!" In an earlier time when taking the promised land of Canaan, the intrepid Caleb boldly told Joshua (Joshua 14: 10-12, my paraphrase), "Give me the walled city high on the mountain *where the giants are*; I am as strong now as I ever was!" (He was eighty-five!) For some, those responses may qualify as at least moderately *toxic* masculinity, a new term recently thrown out by psych doctors, social justice warriors, the homosexual activists, and radical feminists that essentially refers to . . . well, men who act like . . . men. With the obvious dearth of male leadership and the suppression of men in this confused culture, we unquestionably need more real men, good marriages, and noble fathers. I never had the faintest notion I'd write anything related to the subject. But with the now not-to-be-trusted fake news, the negative side of the Internet and social networking, along with the stifling blanket of political correctness on our culture held down by the thought police and the radical feminists, some guidance and truth-telling for confused young men is in order.

Left-leaning academics, radical feminists, Hollywood/entertainment types, and then the advertising industry spearheaded the early attacks on men and manhood years ago. Are you even aware that many universities teach in one form or another that men—especially white men—are the source of virtually all of the world's problems? The put-down and denigration of men can now regularly be seen on television sit-coms, commercials, other types of ads, from caustic comediennes, and even the greeting card industry. It seems that it is politically correct and generally acceptable to *only* abuse and make fun of *men*-mostly white men-as incompetents, buffoons, slothful idiots or worse (something absolutely forbidden today with women or any ethnic minorities). If you've somehow not picked up on this, just pay careful attention to various ads and commercials for a few weeks. For the last few decades men and masculinity have been under serious attack at most colleges and universities, and I knew we'd crossed a dangerous threshold last year when even in the Lone Star state it became official.

I'm referring to Texas, that gave us many of the intrepid Alamo fighters, Captain Leander McNelly and other legendary Texas rangers, Robert E. Lee's fearless "shock troops" (the Texas brigade), victorious WWII Naval Commander Chester Nimitz, Harvey Penick, Chris Kyle, and A. J. Foyt—just to name a few. American masculinity came under official formal attack at the University of Texas in Austin, as conventional historical male behavior— *toxic* masculinity, if you will—was classified **a "health crisis."** It was claimed that all kinds of personal and societal problems unfold when a man acts *like a man*. Of course I'm aware that Austin is a left-leaning, large university city, the seat of a huge state government, and even a *sanctuary* city for lawbreakers, but still . . . a manhood assault outa Texas?! What would Stephen F. Austin, William Travis, or Sam Houston have thought, or anybody else even a few years back? I'll bet Roger Staubach, Earl Campbell, Randy White, Nolan Ryan, and A. J. Foyt are still stunned.

My wife shared a recent copy of a *GQ* magazine, which showed up in our mailbox (a magazine I would never waste my own money on), and said—with sort of a smirk— "you've got to see this." On the cover was this unsmiling, strange-lookin' dude (Pharrell Williams) dressed in a huge, pleated, bulky, yellow single garment that looked like a hybrid of banana peels, yellow squash, old mustard-colored sleeping bags, and looking like a dormant sulfur-covered pyramidal-shaped volcano that seven kids could easily hide under, framed by the title: *The New Masculinity Issue* (Nov., 2019). Gazing at the *GQ* cover, my immediate first thought and question- right or wrong: 1) this guy must have a low self-image, and 2) this dude is going to teach *anybody* about authentic masculinity?! I must admit I didn't recall or recognize Mr. Williams at first, as I'd grown so sour on hip-hop/ rap music the last several decades, in large part because of the depressing, foul, and profane lyrics promoting violence, disrespect of law enforcement, rape, and an overarching focus on sex. That aside, as well as his unique front cover wardrobe selection, the lead-in on the opening page of the *New Masculinity* article states: "ONE THING THAT THE VIOLENT, TOXIC, AND MISOGYNISTIC REVELATIONS OF RECENT YEARS HAVE MADE CLEAR IS THIS: MASCULINITY MUST CHANGE." And then, "WHICH

RAISES THE QUESTION: WHAT SHAPE DOES A *HIGHER* MASCULINITY TAKE?" I forced myself to read the comments of the *GQ*-described *"fearless"* thinkers, activists, and so-called artists who are supposedly "leading the way," in changing and redefining masculinity for the better, including those of the so-called *"EVOLVER"* himself, Pharrell Williams. Forcing my eyes to observe the photographs of some of these profiled individuals was even more challenging. I can honestly say that it was the first time that I ever actually got a bit nauseous reading and *looking at* a magazine article.

There are so many false claims, terms, and statements in these articles I couldn't begin to cover them all, as these people and *GQ* magazine attempt to justify their own lifestyle choices, force them on society, and redefine gender, masculinity, manhood and more. Several times in this current book the importance of language, as well as defining the terms of the debate, is mentioned. Why the sudden popularity of *misogynistic*, for example? This word is often simply defined as "strongly prejudiced against women." Are there men out there who don't respect women like they should, and have definitely dropped the ball as husband, father, or the spiritual leader of their family? Of course there are-always have been. The men I associate with, all heterosexual and mostly married, are *not* "strongly prejudiced against women." In fact, I am regularly inspired and amazed at the self-sacrificing love, service and commitment to their wives by my Christian brethren. It is a beautiful thing to behold.

It is unfortunate yet uncomfortably fascinating to observe what unfolds in society when the moral law, or divine imperative for ethics, is cast aside. The loss of biblical authority in these formerly United States (as well as in Western Europe), was catalyzed by the science dudes (white lab coats) over a century ago, with its unfortunate replacement by materialistic Darwinian evolution. Somewhat ironically, in the just-mentioned *GQ* interview and article on Mr. Pharrell, he is not only heralded as the '*Evolver*'; within just a paragraph and a half (p. 72), "evolve" is mentioned three times, as it talks about his "past missteps and personal evolution." What drivel; evolution is not the goal, evolution is not even reality. The goal is *maturation*, backbone and character development, and a proper walk into authentic manhood. Do not be misled or intimidated by the thought police, psych docs, or the sexual revolutionaries. Readers need to decide for themselves, but the opinions and advice offered in the November, 2019 issue of *GQ* Magazine are *antithetical* to real manhood.

Let's explore true and proper American masculinity and manhood. There is an important verse from I Corinthians 16:13 relative to men and manhood:

"Watch ye, stand fast in the faith, *quit* you like men, be strong."

The old English word *quit* means play . . . *play* the man . . . or *act*, i.e. act like a man, and is immediately followed with *be strong.* The very next scripture (verse 14) adds to "do everything in love." The implication is the reality of a certain properly courageous way for men to behave, or act, relative to their God-given role, hinting at something probably *different* for women, and their role. It takes both men *and* women to "image" Jesus. As mentioned earlier regarding gender, the Bible teaches equality and equal worth of the two, but with different roles. The Creator made us different, but equal, and Scripture commands men to treat their wives as "joint-heirs." As historian, author, and Pastor Harry Reeder often points out, equality is *one's standing*; the same blood of Jesus was shed for a woman as shed for a man, and the same Holy Spirit that indwells a man is in a woman. Although anathema to leftists and the gender-confused, equality does *not* mean interchangeability. It's not a matter of superiority or inferiority; it is just a matter of *distinctiveness.* Let's break this down and explore historical American manhood, which is synonymous with Christian manhood, which the current culture abhors. Very little of the following discussion was an original thought or idea of mine. I learned this from successful men and women from history, the Bible, and the present. They have been inspirational models and mentors-all with *good fruit* on the tree.

Back to the command for Christian men in the first part of I Corinthians 16:13, where it says *"watch ye,"* or "be watchful." This is short but important: Keep your senses sharp and always be alert and watchful, and don't live your life carelessly or preoccupied with frivolous things. Then, *"stand fast in the faith"* (or "stand firm") follows. In other words, don't back off or give ground to the enemy, or what you know is wrong. The next phrase in this all-important manhood verse: *quit ye* (or act) *like men*, which is then followed by the simple but powerful command, *"be strong."* Wow; it is difficult enough to always be strong, but it's even more challenging—and more important—to *finish* strong. There are close to 350 biographical sketches of men and women in the Bible, but only 67 finish strong. (One thing about the Bible; it's *honest*, and constantly points out man's sin, weaknesses, and foibles). Be a good finisher. Before we hash out what it really means to act (*quit ye*) like men, recall the following verse: "do everything in love." From I Corinthians 16:13-14, these two precise life-action verses for men can be summarized: *Be strong and courageous to embrace the responsibilities of life, and sensitive and compassionate in the personal and family relationships of life.* Put another way: be *lion-like* to embrace the responsibilities of life, and *lamb-like* in the personal and family relationships of life. Mimic Jesus: be a lion and a lamb.

In this current dysphoric culture, how does a real man *stand* his ground? *"Quit* ye like men," or "act like men"; what does that really mean? There are many different and questionable messages being sent today about men and manhood. How is a real man supposed to act? It's all about character.

Here are five character traits to consider (all modeled by the aforementioned heroes Lee, Jackson, and Chamberlain) that are found in real men of character and many rock-solid dads and American heroes: 1) Dependability, 2) Integrity, 3) Civility, 4) Sexual purity, and 5) Humility. As we explore this briefly, it is admitted this view is not politically correct or favored by academics, radical feminists, homosexual activists, the mainstream media, and the secular left. They deny the *distinctiveness* of real Christian manhood, and they sadly deny *its value*. A man that dares to live by these five character traits today is swimming hard against the current, but it's the right thing to do. We need such men in times like this. Courageous men with backbone have a character dynamic that manifests the following principle: *Circumstances do not dictate your character; they reveal it, and present the opportunity to refine it.*[49] Again, this goes against the mindset and culture of today, which often makes these claims: *Oh, you couldn't help it, it's a genetic problem*; or, *it's not really your fault, you're the product of your environment*. Or, *she couldn't help it because of her parent's troubled relationship, or the rough neighborhood*. Do not misunderstand; your environment and other factors are important and can have an effect, but one is not automatically the product of their environment. Conversely, *we* produce— or *can* produce—our environments, and make decisions relative to how we're going to respond to them. In other words, when the going is tough, are you going to be *overwhelmed*, or an *overcomer*? Consider just two of America's overcomer heroes already mentioned. Robert E. Lee was *abandoned* by his father at age six, with the family left saddled with debt, and he never saw him again. Thomas (later Stonewall) Jackson was orphaned as a young little boy in rough western Virginia, and had just a year of school education, until surprisingly accepted to the U.S. Military Academy at West Point on a last-minute miracle. Thrown behind the eight-ball at such a young age, and later dealing with tragedies and challenges with family and career, it is more than impressive what these two men accomplished. The Apostle Paul would likely have exclaimed, *Hupernikao*, "more than conquerors."

Jesus, among many other monikers from scripture, is described as "the lion and the lamb." This is a great model for real manhood, symbolized as a mighty lion, or a warrior. Examples include: "The Lord is a man of war," (Exodus 15:3); "casting the moneychangers and merchants out of the temple of God," (Matt. 21:12); and boldly telling the corrupt Hebrew Pharisees (who were plotting to kill him) in John 8: 44, probably with their mob and

---

[49] Dr. Harry L. Reeder, III, from numerous discussions and seminars while walking America's Civil War battlefields; also found on the 6 CD Set *Christian Manhood Illustrated*, Briarwood Presbyterian Church, Briarwood Bookstore, Birmingham, AL, 2009

bodyguards present, "Ye are of your father, the devil, and the lusts of your father ye will do. He was a murderer from the beginning"! Wow, this is the very definition of *audacious*. Consequently, "the scribes and chief priests . . . *feared* him" (Mark 11:18). Everyone needs models; perhaps Jesus' strong lion-like model should be ours in fulfilling life's duties and responsibilities, standing for right and honesty, and overseeing and protecting one's family. This image contrasts greatly with some of the contrived paintings and pictures of a pale, skinny, long-haired, effeminate Jesus with a probable wimpy, soft damp handshake one often sees hanging in some church hallways, does it not? Understand that neither age nor size nor good looks makes the man— *it is his willingness to accept responsibility*—to merely *act* like a man.

At the same time, a man should be sensitive and gentle, or *lamb-like*, in personal and family relationships . . . a lion *and* a lamb. Remember the following verse (1 Cor. 16: 14) says to do everything *in love*. Could this be considered "toxic" masculinity? One of my most memorable and exciting reads ever comes from J. A. Hunter, one of the most respected and admired of the African big-game professional hunters and game wardens of the 20th century, in his exciting autobiographical page-turner called *Hunter*. In his profession he faced many perilous situations with the men and dangerous beasts he dealt with: hunters (some wealthy, some not), photographers, trackers, and gun-bearers—where their courage (or lack of)—and character were exposed, often in life-threatening situations. Sometimes the hunter became the hunted. The statement that surprisingly stuck with me, though, from that unforgettable real-life adventure book was the claim by Mr. Hunter that *he never met a truly brave man that wasn't kind*. Kindness, or being sensitive, gentle, and compassionate in personal and family relationships, is a key character trait in real Christian men. Did you ever wonder where the term *gentleman* came from? It hearkens back to our truly Christian era, when a strong and courageous, but *gentle*, man was prized and honored; it's far more than a name on a non-gender-confused bathroom door.

Let's consider those manly character traits, starting with dependability. 1) *Dependability*: you say what you mean and your word is your bond, and when things get tough you'll be there. You do what's necessary even when it's uncomfortable. You are persistent and consistent. Your wife and children can count on you; you're rock solid. As a Christian, you "walk in a manner worthy of your calling." Your walk manifests your salvation. 2) *Integrity*: your character: you treat and act the same around everyone, and it's reflected in how you act and what you do when no one is looking. Whenever possible, you look to do the next best thing. You speak truth; you mean what you say. 3) *Civility*: has to do with the words you use, and the way you speak to and about others; and the way you treat people. One of Stonewall Jackson's maxims: "Avoid triumphing over an antagonist" (don't talk trash, mock, or taunt). The great Booker T. Washington said this about Robert E. Lee: "No one could live in the presence of Robert E. Lee and not be the best." Also concerning Lee's civility, let's go back to the early morning in December, 1862, up on the sunny heights above Fredericksburg, with the Union troops about to attack (and get decimated). One of Lee's generals claimed he prayed the night before "that the enemy would charge over, and he and his men would kill them all, and they go to hell!" Readers should know that Lee would not tolerate profanity, and also could never bring himself to call the opposing soldiers "the enemy," or "Yankees," or something worse. He always referred to them as "those people," or "Mr. Lincoln's people," or late in the war "Mr. Grant's people." Lowering his field glasses, Lee eyed the outspoken general and said this: "General, I prayed for those people also, and I asked God to save their souls and that they would all just go home, and leave me alone."

A final stellar illustration of duty, finishing well, and the character trait of dependability took place near General Lee's headquarters tent in the woods at Appomattox in early April, 1865. Lee's depleted, half-starved army is down to 12-18 thousand men. A courageous last-gasp attack led by John Brown Gordon breaks through, only to discover many thousands of fresh awaiting federal troops as he crests the hill. Lee's army is now surrounded on three sides. Word is sent back to Lee; there are no re-enforcements. General Lee has few options, and asks General Henry Wise, "What will the country think, if I surrender?" Wise responded: "General Lee, you *are* the country. These men have only been here for you the last two years." Lee then spoke, "I would rather die a thousand deaths, but I have often said these matters should never have been committed to a military solution. But we did, and now God in His providence has spoken. It is my *duty* to surrender." About that time the sharp young artillery leader, General Edward Porter Alexander, spoke up as he pointed toward the distant Blue Ridge: "General Lee, the west is yet open. We can break into bands and guerilla warfare. We can terrorize the enemy until they give us our independence. The men will do that for you!" Looking at his intrepid artillery officer, Lee spoke: "Young man, bushwhacking may be compatible with your views, but if I were to do what you say, what would happen to our wives and children? We have standing armies in the field. Look what is happening now (as he alluded to the North's *total war, scorched earth* policy). No, it is our *duty* to surrender." Edward Porter Alexander later said, "I realized at that moment, that I was standing in the shadow of a man who thought much grander thoughts than I."

Duty, dependability, character; they were all watchwords for Robert E. Lee, and they should be for us. He had previously said that duty is something a man need not fall short of, and never apologizes for as he embraces it. This idea is not popular in today's secular culture with so much *self*-focus. This theme continued shortly thereafter at the surrender meeting with General Grant, as Lee's concern was centered on his men. Lee wanted to do the right thing, and he dreaded the possibility of his men going to northern prisons, where there was a 60% mortality rate. He asked Grant for two concessions, after the northern commander generously granted parole to Lee's men: 1) Can I have food for my men? And 2) Can they keep their horses (there was still time to put in a spring crop)? Lee was quietly elated; Grant agreed to both.

Incidentally, but back to character, when General Lee stands up to leave the McLean house after agreeing to surrender terms, everyone else stands up (the room was half-full of Union officers). Then when he mounts and leaves, with only George Tucker—his sergeant orderly—and his assistant Colonel Marshall, fourteen Union Generals outside stand silently in awed respect and *salute*. Just amazing, when you think about it; Lee was the loser, not the victor! A succinct summary of the character traits of dependability, integrity, and civility: *Say what you mean, mean what you say, and don't be mean when you say it.*[50]

The two remaining character traits of real manhood are *moral purity* and *humility*. Moral purity is really very straightforward: We are to flee immorality, and pursue the enjoyment of sexual intimacy within the bounds of marriage with our wife, a beautiful gift God has given us. Try to imagine the lives of Generals Robert E. Lee, Stonewall Jackson, and Joshua Chamberlain. These men were flat-out heroes, chased by women regularly (yes, even then), and there were numerous incidents, and of course opportunities. Lee was still stout and well-built at age 54 when the war started; he was earlier called "the most handsome man in America." He spent 50-60% of his military life away from his wife on post, and greatly enjoyed being around women, but with never a hint of sexual impropriety. Stonewall Jackson, a living legend the last two years of his life, happily married and the dedicated Presbyterian, was embarrassed by female pursuit, and always took great care to make sure he was in the company of others when women were around. The same can be said of the Union hero of Little Round Top, Joshua Chamberlain, and his poignant letters to his wife Fannie are worth reading. These were men of integrity, and followed the Bible's precepts and program: chastity before marriage, sexual purity during marriage, and within the bounds of marriage wonderful sexual enjoyment.

Humility is the final character trait of real men and great leaders. Ephesians 4:2 says to walk "With all lowliness and meekness, with longsuffering, forbearing one another in love," or, "with all humility and patience, showing tolerance for one another, in love." Today's macho men with big egos and no humility are usually far removed from real manhood. Many

---

[50] Reeder, Ibid, battlefield chats

examples of humility can be found in the life of General Joshua Chamberlain, who decided to take leave from Bowdoin College (in Maine) and academia, with no formal military training to serve and do his part to save the Union. A strong Christian, highly educated, and somewhat proficient in nine languages, he was not the typical volunteer. Although offered a Colonelcy by the Governor, Chamberlain—with humility—said something like this: "Sir, I have had just a modicum of military training, and while I am humbled by your partiality, it would not be fair for the men under my command, to be commanded by me. I would accept a Lieutenant-Colonelcy, if you could put me under a West Point-trained Colonel, to learn."[51] So night after night, well into the night, Chamberlain eagerly humbled himself and thankfully trained under Colonel Adelbert Ames, studying, training, drilling, and learning. Little did they know that both of them would eventually be awarded the American military's Medal of Honor.

When the first battle finally came, Chamberlain was ready. Seven months later the extra night-time training, his humble spirit, and ever-learning attitude all came to fruition. It was on a sweltering summer day on July 2nd, 1863, at a rocky hill called Little Round Top near Gettysburg, Pennsylvania. He was hastily placed there with his regiment by Colonel Strong Vincent, who would take a bullet thirty minutes later that would cost him his life. Positioned there to anchor the critical left end of the Union line on Cemetery Ridge, now *full* Colonel Joshua Chamberlain, he and his tough 20th Maine Regiment withstood a minimum of seven Confederate assaults. After performing two amazing military maneuvers without giving ground—lengthening or "refusing" his line (to avoid being flanked), and then—when out of ammunition—making an unheard-of left wheel bayonet charge, the flank was secured and the day was won. It was a singular military movement in the all-important 3-day battle, perhaps one of the most important in the entire war. I share General Chamberlain's following remarks, made many years later on a return to the famous field of valor, as they illustrate not only what can come from *humility*, but how the Christian *character* of a real man is developed and revealed. When asked *how* he and his exhausted men were able to hold out and perform that audacious,

---

[51] Please consider the following books to learn more about the life, character, and military career of Professor, General, Governor, and Senator Joshua Chamberlain: 1) Alice Rains Trulock, *In the Hands of Providence*, The University of North Carolina Press, 1992; 2) Joshua Lawrence Chamberlain, "*Bayonet! Forward*," Stan Clark Military Books, Gettysburg, PA, 1994; and 3) Joshua Lawrence Chamberlain, *The Passing of the Armies*, G. P. Putnam's Sons, 1915, and Bantam Books, 1993

heroic maneuver at Little Round Top, culminating in a fearsome bayonet charge after running out of ammunition, he shared this with his audience:

> We know not of the future, and cannot plan for it much, but we can hold our spirits and our bodies so pure and high, we may cherish such thoughts and such ideals, and dream such dreams of lofty purpose, that we *can know* what manor of men we will be whenever and wherever the hour strikes, that calls to noble action. No man becomes suddenly different from his habit and cherished thought.[52]

In other words, in moments of crisis, men will do what have become the habits of their heart. At that decisive moment, what you have held in your thought life and your heart is what comes out. Courage is not the absence of fear; but the willingness to do one's duty in the face of fear. How about you men out there? The storms and crises are going to come. Our lives are dictated partially by events beyond our control, so control what you *can* control: how you spend your time, what you read and watch, care of your body, who you associate with, etc. We're not all military warriors, but a crisis or tsunami is going to come, likely more than once. Be ready; work on those heart habits and thought life. Remember, circumstances don't dictate your character, they reveal it, and present opportunities to grow and refine it. Leadership is a fascinating study. Great leaders come from both *nature* and *nurture*. Some, it occasionally seems, are *born* to lead (the DNA/heredity factor). Most are *made* (nurture/environment); leadership is in large part *learned*.

There is an incredible scene around the throne of God in heaven, described by the Apostle John, as he was translated there in the spirit to be shown "things which must be hereafter" (Revelation, Chapter 4). Revelation is the only book of the Bible which *promises* a special blessing to those who read it, and even to those fortunate enough to *hear* it read (see Rev. 1:3). There are many issues and words of importance in chapters 4 and 5 of the Book of Revelation, but my emphasis from there is the picture of Jesus himself as a wonderful manhood model—a lion *and* a lamb—*strong and courageous*, yet *sensitive and compassionate* (or lamb-like) when called for.

A quick return to a month in the life of Robert E. Lee illustrates this courage-kindness dichotomy of real manhood, and you can formulate your own opinion about the *toxic* part. A loss of manpower, illness among the

---

[52] Joshua Lawrence Chamberlain, *"Bayonet! Forward,"* Stan Clark Military Books, Gettysburg, PA, 1994, p. 188

troops, and lack of food, along with ill-fed and diminished numbers of horses plagued General Robert E. Lee and his vaunted citizen Army of Northern Virginia as the dogwoods bloomed in the spring of 1864. Lee's rag-tag army prepared for the onslaught from across the Rapidan fords by the powerful Army of the Potomac, now under the overall command of General Ulysses S. Grant, who was finally just days away from going face-to-face, for the first time, with the bold Confederate commander. Aware of Lee's diminishing manpower and other challenges, with confidence in his own rested and well-equipped massive force, Grant was quietly confident of a smashing quick victory. In the springtime the large armies, with all of their artillery, wagons and supplies, as well as men and horses, had to wait for the abused and wet muddy roads of late winter to dry up before they could mobilize and renew the slaughter. In the first week of May, General Grant was literally chompin' at the bit—actually smokin' and chompin' on dozens of cigars—to attack with his superior force (120,000). When he was finally able to make his move a terrible 2-day battle commenced in and around the second-growth woods not far from the old Chancellorsville site-The Battle of the Wilderness. One letter from a northern infantryman back to his family shortly after this battle claimed the battle of Gettsyburg (the previous July) was a just a "skirmish" compared to the Wilderness fight.

At a critical moment when it appeared Lee's army would be overrun by many thousands of northern troops, he rallied a mere 100 to 200 hard-core fighters and prepared to lead the charge himself, directly across an open field against perhaps the toughest unit in the Union Army—the "Black Hat boys" from the northcentral states—also known as the "Iron Brigade." They were legendary, even before the war was over. He quickly called his small group into line as he prepared to lead them in a counter charge against thousands of battle-hardened men. The questionable assault prepared to begin with Lee right behind the few men on his beloved horse Traveler. General Lee refused to heed their calls to go back to the rear. Only when a Colonel spurred up close and shouted in his ear that Longstreet was finally up with his men—and needed quick orders—did Lee reluctantly give way to the leading group of tough Texans. A small understated marker highlights this focal point of courage and stalwart American manhood at the Wilderness battlefield to this day; hundreds of Hood's Texans knowingly charged to their death . . . but saved the day (17,500 Union casualties in the 2-day battle) for Lee's army. The same thing happened a week or so later as the battle shifted quickly to Spotsylvania, where the courageous Union soldiers threatened to split the entire southern line (at the salient, or *Muleshoe*) and turn it into a rout. Lee once again, this time with lead flying everywhere, attempted to lead the charge on his fine steed to save the day, as his loyal men shouted "Lee to the rear, Lee to the rear!" Only the efforts of John Brown Gordon and a tough sergeant who grabbed Traveler's bridle stopped him. [53] Two weeks later at Cold Harbor, as the non-stop fighting continued, Lee once again wouldn't budge in the face of danger, ignoring more calls of "Lee, go to the rear!". . . and then watched some 23,000 men go down in a twenty minute bloodbath in an ill-advised charge ordered by General Grant . . . and the Confederate leader never blinked or wavered. It's called *cool under fire*. The North and South had been fighting for three years at that point.

From May 5th to June 3rd, 1864, for 28 brutal days of constant fighting, men of courage on both sides fighting for what they believed in, from the Wilderness to Spotsylvania Court House to North Anna and then Cold Harbor, General Robert E. Lee and his ragamuffin army, with essentially half the men of Grant's powerful force, inflicted more casualties on the northern Army of the Potomac than the number of men *Lee had in his army*. Illustrated in this titanic sanguinary struggle in May and June, 1864— exhaustive non-stop fighting for a month or more— by undermanned but not overwhelmed General Robert E. Lee, was true warrior grit and *lion-like* behavior, to fulfill one's duties and responsibilities (as he saw it), as a leader of men (*strong and courageous*, I Cor. 16:13) .

How about *the lamb* part? Real manhood is not about macho, profanity, shouting and drinking, but calls for sensitivity, gentleness (lamblike behavior), *and love* in personal and family relationships (1 Cor. 16: 14, "*everything with love*"). This is poignantly illustrated in the loving care, as a young West Point Cadet, Lee lavished on his invalid mother as he took some time off to care for her, and then years later the same thing for his invalid wife, Mary Anna (many feel she likely had what today we call severe rheumatoid arthritis, or possibly fibromyalgia); she became wheelchair-bound before 60. Lee, one of the most capable and effective soldiers ever in the history of the U. S. Army Corps of Engineers, was ahead of the times; he modified their house to be *handicapped accessible*, and pushed her on strolls whenever she felt up to it. Lee's love and playfulness with children, even those not his own, is documented and was well-known to all of the residents around Arlington and West Point, as well as in Lexington, Va. during his last five years after the Civil War. *Lamblike*—gentle and sensitive, with all things done in love—in personal and family relation- ships, is surprisingly to some, an important part of *manning up*. Gentleness, or kindness, is one of the nine-fold fruits of the spirit of a real Christian. An anecdotal story showing the lion *and* lamb-like traits (or heart) of General Lee took place late in the war at Petersburg, when he ordered his men in a sud- denly dangerous spot to quickly get under cover from their exposed position

---

[53] James M. McPherson, *Battle Cry of Freedom*, (New York: Ballantine Books, 1988), 209-210.

and a withering enemy fire, and then Lee exposed himself to further risk by hustling out to pick up a baby bird that had fallen, and returning it to the nest.

These real manhood characteristics just discussed have been manifested by countless heroic fathers and American heroes-many known to history, many more unknown. Robert E. Lee is one of my favorites; I could just as easily have talked about eight or nine others. I probably could not have made it in life without some heroes—or *models*—and a few mentors. **Mentors** are those who are still alive and can guide you; I call them 'life coaches,' and they can help you with (as Baloo would say) the *vicissitudes of life*. English professor and Christian writer C. S. Lewis pointed out that leaders and successful people tend to have or find mentors, and build relationships or friendships with such individuals, who can help them *recognize and see beyond their own limitations*. Don't make these valuable mentor-friends your models (or heroes) yet, while they're still alive—and are still *fallen* human beings—they can still mess up, and the jury's still out on them. All young men and women do need to get several inspirational models. Get them from history, or the Bible—the jury *is in* on them—their lives can be fully examined.

Does this picture just presented qualify as *toxic* masculinity? I'm not really sure . . . I just know America now needs more of such backbone and true manhood than ever before. The following quote from one such dude, President Theodore Roosevelt, has been used a lot, but it is apropos here:

> It is not the critic who counts, not the man who points out how the strong man stumbled or where the doer of deeds could have done them better. The credit belongs to the man in the arena, whose face is marred by dust and sweat and blood, who strives valiantly . . . who knows the great enthusiasms, the great devotions, who spends himself in a worthy cause, who at the best knows in the end the triumph of high achievement, and who at the worst, if he fails, at least fails while daring greatly, so that his place shall never be with those cold and timid souls who have never known neither victory nor defeat.[54]

We need more millennial and younger men and women *in the arena* today. What a great commentary, by perhaps our most eclectic and larger-than-life president; he probably would have been a starter on my 21st century toxic masculinity all-star team. Other probable candidates for that team

---

[54] Thomas Russell, ed., *Life and Work of Theodore Roosevelt*, L. H. Walter, New York, p. 257

are a group of impressive young men often at the point of the spear I had the privilege of spending time with recently in southwest Florida-some special forces warriors of the United States Air Force. These dudes laugh at college "safe places" and people living by just their *feelings*. Their motto: *Adversity Breeds Resilience*. Our troubled young men in this cold civil war need to grab on to that one . . . and, no matter what the psyche docs or liberal academics say . . . they need to finally *man up*.

9

THE *"WEE ICE MON"* ... HE *EARNED* IT ... HOW ABOUT YOU?

NOTHING SUBSTANTIVE, NOTHING meaningful, nothing lasting and solid comes without hard work. This includes study and learning, listening, practice, finding mentors, sweat, going early and staying late, serving, going the extra mile, and what you do after you've done everything you're supposed to do. In the end, you've got to believe in yourself, that you've done the work, that you've prepared well and been true to yourself, and that you can succeed and perform under pressure-even *thrive* under pressure. Because you've sacrificed and paid the price, you've given your all, you've done your homework—and you are ready. You have *earned* it, and if opportunity arises, you deserve it—you need to feel this way deep in your heart—a grounded, in-your-core earned confidence.

Consider one of the greatest golfers of all time, hardscrabble Texan Ben Hogan. Starting in the dusty days of early Texas golf in the 1920s as a kid working (sometimes fighting) his way up through the caddie ranks, Hogan later practiced and sweated, worked and experimented-whenever he could-on the practice range day after day, often in the oppressive Texas heat. His goal was to not only develop a great golf swing, but one that would hold up under pressure, *even thrive* under pressure. He kept detailed notebooks with him on the practice range, experimenting and grinding (with no swing coach, no self-confidence guru, definitely no gender dysphoria counselor, or business manager) by himself, trying to find that repeatable, dependable swing. The tough little Texan labored over eight years on the pro golf tour, pinching pennies most of the way, as he developed his singular machine-like swing, paying the price before he won his first tournament. But after that grinding, long paying-his-dues period-which ran into the second world war-Mr. Hogan finally won his first tour tournament, and then it was flat-out *Katie bar the door*! From 1945 to 1953 the diminutive golfer from Dublin, Texas won over 50 tournaments and was at times unbeatable, dominating professional golf after WW2 in profound fashion. In 1953, four years after a terrible auto accident at the peak of his career, when many thought he would never play golf again because of serious injuries, Mr. Hogan played in only six tournaments, won five of them, including the three major championships he entered—the Masters, United States Open, and British Open—and set record low scoring records in each one . . . total domination. For you linksters, Hogan apparently declined to enter the PGA Championship that year due to the unique but grueling 36-hole match-play final, not confident his ailing legs—from the past auto accident—would be able to handle it. In his one and only British Open appearance, which was that year at Carnoustie, Scotland, where he not only won but set a record score, the Scots dubbed him "the wee Ice Mon." Even his top rival Sam Snead, with great respect, once said Hogan was "like cold steel." Incidentally, the gritty Texan was not that enamored with the wild grass of Carnoustie. Shortly after his win, at a celebratory function back in the states, the salty champ said this: "I've got a lawn-mower at home; I'll send it over there."

As an interesting sidebar for you golfers, one of the things Mr. Hogan did in preparation for his one British Open appearance was walking the entire course backwards, starting at #18, with detailed note-taking. The Hogan mindset reminds me of a military friend's mantra, "be prepared, not scared." The *Wizard of Westwood*, legendary UCLA basketball coach John Wooden, once said to study, prepare, prioritize and practice so that you have a great chance of "being at your best when your best is needed." Sounds like Mr. Hogan; I wonder if they knew each other (they were only two years apart in age). The SAS (Special Air Service) guys—the elite special forces of the British (some tough dudes- often *toxic*, and American allies)—have a creed: "train hard, fight easy."

In those dominating years from 1945 to 1955, constantly dong battle on the links with some of the all-time greats like Snead and Byron Nelson (all three born in 1912), I strongly suspect Mr. Ben Hogan, in his heart of hearts, had that core confidence and felt like he'd paid the price-he'd earned it. How about you?

Regrettably, it seems more young men today are soft, wimpy, confused, victims of political correctness and left-leaning secular education, lazy, and have an anathema to the hard work ethic that millions of Americans historically embraced to escape poverty and pursue the American dream. Physical labor, sweat, hard work, and even occasionally getting your fingernails dirty was formerly considered good for physical and spiritual health and the suppression of pity parties, not to mention your long-term future. I'm reminded of the hero in Winston Churchill's only novel *Savrola,* when he asks "Would you rise in the world?" Then, "You must work while others amuse themselves." Some of my construction and mechanical friends tell me it's difficult to get good young craftsmen and skilled laborers-nobody wants to work hard anymore.

I heard a report last month where it was discussed how young men want the success, the condo or nice home, and other material things but don't want to do the hard work required to make the bucks to get it. I was always impressed that best-selling American western writers Louis L'Amour, Zane Grey, and Elmer Kelton touched on the theme of hard work and the benefits of physical labor quite often, but sadly their wonderful books are—of course—unknown to today's coffeehouse misanthropes, the gender-confused and the video-game and/or porn-addicted. Surprising to many, Zane Grey, perhaps the best-selling author in North America the first half of the twentieth century, always referred to his popular rugged western novels as *romances,* not westerns. There was always a love story involving a man and a woman, the triumph of right, and that formerly internationally-recognized tough, never-quit American manhood. None, by the way, included gay couples. None had a sense of entitlement, or the idea that the world owed them something. They're just great accurate history, great adventure, and simple but super reading. These three classic western writers were an inspiration to many young men (like the old John Wayne movies and TV westerns) and women with their American values and real history, and still could and should be today. They are probably not on the recommended reading lists of the politically correct, the radical feminist movement (whose participants are not very feminine), homosexual activists or the social utopian multiculturalists. But they are still wonderful and inspirational, for men and women of all ages . . . but perhaps especially for adventuresome *young* men with . . . dare I say it . . . a toxic masculinity gene.

10

WHAT'S YOUR "ADVERSITY" SCORE?

RECENTLY A LIBERAL cone-head seriously proposed changes in the scholastic SAT test and its scoring by adding and factoring in one's so-called *adversity* score, based on the teenager's answers to questions relative to his or her family, socioeconomic situation and life environment (specifically with an emphasis on urban African-Americans). It is just another example of political correctness/identity politics/social *justice* run amok; one of the major goals previously with the SAT test was being objective—no favoritism—and assessing real knowledge. When I first heard about this proposed nonsense earlier this year, I immediately thought about those middle and lower-middle class Asian-Americans, who we all should be so proud of and impressed with, with their family values and emphasis on education. They do well in academics and on the SAT test (as do some other groups, too; not necessarily white Caucasians). Like other leftist proposals, this newest one is senseless, and again hints of racism.

Perhaps I can illustrate my point best using a real-life example with Tim, one of my best friends. Tim is a white male in his mid-fifties who grew up poor in the northern North Carolina mountains, and currently has a very successful business in finance and insurance in the greater Charlotte, N. C. area. We have a bond of not only both being of mountain folk from the Appalachians, but a love of hiking, wildlife and trout fishing, as well as sports and holistic health, along with biblical truth.

Tim grew up in a house with no indoor bathroom, a coal stove in the little house for heat, with a woodstove to cook meals on, drank unpasteurized milk from the cow out back which he himself often milked (at age 5-6), and bathed in a metal tub. He split and stacked wood regularly (elementary school age), got reduced-price lunch in school, worked at age eleven putting seat bottoms in chairs and other manual labor jobs, and never knew his biological father. Especially you young men and women, understand that life often isn't fair; but keep learning and doing what you're doing, and simply do the best you can-you can do no more. Even the apostle Paul, in referring to life in this fallen world, describes it as one of "groaning" and "travail" (Romans 8:22). Tim regularly had to take the "poop" bucket outside, feed the chickens, carefully collect the eggs, and other chores. He later worked hard on the mountain country Christmas tree farms, where he was exposed to glyphosate (Roundup) repeatedly (without safeguards), did other odd jobs, and somehow graduated from high school while becoming a high school sports star. He subsequently got his college degree while getting married to his high school sweetheart and busted his tail to get where he is today. Tim is unquestionably the type of guy you'd want beside you in a foxhole. Right after the *adversity* score nonsense came out a month or two ago, he texted me about this and wondered if his adversity score would have gotten him into Harvard (with a big LOL!). As Pastor/historian Dr. Harry Reeder stresses, "circumstances do not determine your character; they reveal it, and present opportunities to grow and refine it."[55] Tim is a positive example of this.

Then there is my 90-year old friend Neil, who is a patient of mine who I now assist working in his pretty serious garden (yep, I still work for food). Neil was a middle child born in 1929, had one brother and one sister, and while still very young the three of them and their mother were abandoned in Florida by Neil's father, in the tough Depression years of the early 1930s, when money was scarce nearly everywhere. Somehow they found their way back to their home area in the piedmont farm country of Union County, N. C., to scratch out a living on a small cotton farm. There were two functional cotton gins operating in nearby Waxhaw, N. C. then, and Neil shared with me that he was saved at a revival in a Methodist Church there when he was 13 years old. Neil has no memories of his father (being just 2-3 years old when he disappeared), but claims that his mother was a very modest, morally strong woman. They just all chipped in and did what they had to do, with an outside toilet, a backyard well and a flimsy small farmhouse with no insulation, where you "could throw a cat through the wall in some places," according to Neil. Their nocturnal wandering tomcat could always find his way back into the house, even in winter, despite the door and windows being closed. They had some occasional help from a generous farm neighbor or two who lived several miles away. They walked everywhere, and on one Sunday afternoon five-mile walk to visit a relative, his mother finally relented and briefly responded to the ten-year old lad's repeated question, about what happened to his dad. She had previously rebuffed these questions and refused to talk about it, but on that afternoon excursion—perhaps sensing the void in her son's heart—she finally shared that Neil's father got a girl pregnant and had to go away. Wow, what a tough conversation for an abandoned mother of three to finally have, on a dusty Carolina farm lane, with her aching 10-year old son.

---

[55] Reeder, ibid, battlefield manhood/history chats

Neil later dropped out of high school to work and from age 16 to 18 worked in a textile mill in the spinning room as a "doffer." He made enough and saved to where he was able to buy his mother an almost magical Kenmore Sears washing machine when he was 17. At age 18, finally making the borderline-exorbitant salary of $1.62 an hour, he pinched and saved and bought his mother a brand new Crosley Shelvador Refrigerator (the first one with shelves inside the door, by golly!), before he went off to join the navy. Neil served with honor in the navy as a hospital corpsman (pronounced kor'mun, not "<u>corpse'</u>man," as former President Barack Obama prefers) and chief petty officer for twenty years, with four of those with the 2nd Battalion, 5th Marine Regiment, 2nd Marine Division (Chesty Puller's group). He then got out because he wanted to raise his kids in North Carolina. On his return with family, he worked on a framing crew for a while before embarking on a very successful 27-year career with Farm Bureau Insurance, was always involved with his church, and now years later comfortably retired with a batch of grown grandchildren. Significantly— with both Tim and Neil—neither one knew or remembered their biological father, and grew up under less than ideal circumstances. Studies unequivocally show that kids generally do better growing up in a household where mother and father are both present and stay together. Marriages just don't always work out that way-they also didn't for the parents of Robert E. Lee and Stonewall Jackson. The point bears repeating: when things aren't quite perfect—maybe not even close—are you going to be overwhelmed, or an *overcomer*? Tim and Neil had backbone—they were *overcomers*—and lived and continue to live as real men. I am proud to have them as friends. They have been blessings in my life and many others. Is it even possible to develop character without adversity?

Please consider all angles and ramifications before endorsement of this politically-correct social justice hogwash . . . SAT "adversity" score my fanny! On a side tangent, back in the early 1960s after a push by the secular left and under the liberal Earl Warren Supreme Court, the Bible and prayer were officially banned from the public schools, in two successive years, as a secular evolution-based academic and judicial philosophy ("legal positivism") sadly descended across the fruited plain. Significantly, and unquestionably to the *surprise of* the leftist academics, SAT scores immediately began *a 17-year decline*. This had **never** happened before in the previous forty or so years; SAT scores had minor ebbs and flows but basically had always stayed the same. They finally stabilized and started edging up again, due to the subsequent mini-explosion of Christian home-schools and private schools, after the 17-year fall.

Instead of just shaking their heads and getting frustrated, concerned citizens today need to likewise take action or get involved—at least at the local level—in righting this wacky culture of political correctness, and stand for what works, what's true, and real manhood. It is literally getting beyond ridiculous. The Anti-Defamation League, for example, just came out strongly against the use of the well-recognized, cross-cultural *okay*, or *consent* signal, of raised hand with thumb and index finger tips together (forming a small circle), and the other fingers somewhat extended, as they claim it's now a powerful sign of white racism. I don't care if a radical, or even some misguided, no-hope white supremacist somewhere tried to redefine the sign for their own local perverted purposes—this is silly. Have you ever wondered how unhappy some of the activists on the far left must be? It is a somewhat interesting question from a medical point of view (my field), as many have been found to be neurotic or with mental problems. Some, in fact, are aware of their condition, yet seem strangely proud of it, as if it shows their unwavering commitment to *the cause*, such as going all-out to save a tiny rare crustacean that nobody's ever heard of, to block construction of, say, a hydroelectric power plant that would provide energy to businesses and thousands of people. Or, maybe it's those progeny of Al Gore who unjustifiably and vehemently promote imaginary, catastrophic man-made climate change that is soon to submerge and destroy our coastal cities.

Back to adversity scores, growing up in a less than perfect environment, and helping other *image-bearers*. Give yourself away, especially to a young person who has been wounded, or be a mentor, and if you have something you don't need give it to someone who does. As a man's man- Booker T. Washington-used to say, "If you want to lift yourself up, lift someone else up." Do such misguided liberals even realize the message they're sending to minority ethnic groups with proposals like this? It is subtle elitism and borderline racism as well as another manifestation of the dumbing-down of American education. The larger point for young men and women: when things aren't perfect, when challenges abound, when one parent is gone, are you going to be overwhelmed, or are you going to deal with it and move forward? I am inspired by the great George Washington Carver, who came from humble beginnings—a first generation free slave—and developed 332 uses of the peanut and probably saved the economies of four or five southern states with his agricultural research on the sweet potato, the peanut, and other crops, while inspired by the Holy Scriptures. What a sterling example of American manhood, along with his co-worker and educator, Booker T. Washington. These guys were heroic overcomers. These two American heroes were real men of character, and worthy models for young men and women today (but not favored by the left-they were too Christian). Even Jesus speaks highly of overcomers in Scripture. I sincerely hope you young men and women out there develop an anathema to coddling, quit wimpin' around, man-up or woman-up, and choose the *overcomer* option . . . while building and manifesting character . . . and not worrying much about what other people think. Inspirational Hall of Fame Coach John Wooden put it this way: "Be more concerned with your character than your reputation. Your character is what you *really* are, while your reputation is merely what others *think* you are." And remember, don't go lookin' for it, but adversity can be a blessing.

11

THE YOUNG MAN PLAN

IT HAS OFTEN been said that one needs three things for a good life with contentment and confidence: 1-something to do, 2-someone to love, and 3-somethng to look forward to (hope). Hope is so important; when you have it, it gives energy and purpose, and *stick-to-it-ness* in the present. We already know that hard work, courage and character are all involved. In looking around our society these days, especially at young men and fathers, I wonder about their hope and heart habits. We need more young men with old-school grit. How about you? How are you spending your time? Are you guarding your heart? It is absolutely a key to character and backbone development, while developing a sense of quiet inner confidence, being productive, and the three things mentioned above. Let's hash this out a bit.

It is important to be ever-learning. Good self-education is traveling from an unconscious to conscious awareness of one's own ignorance. Even Scripture commands "give attention to reading"; this is part of the biblical command to not only love God with all of your heart and soul, but also with *your mind*. In my studies on military command, looking at some of the great generals and leaders of men, it often becomes apparent that such men are not know-it-alls; these great men just *want* to know it all. Even Mark Cuban, the well-known Dallas Mavericks basketball team owner, self-made computer/high-tech business guru billionaire and entrepreneur, also of TV's "Shark Tank" fame, claims his key to success was *reading*—self-education—two to three hours per day, which he continues to do even now. Mr. Cuban said that it gave him a level of comfort and confidence in his business. Be selective; it's so easy to go off on tangents reading false or non-substantive stuff. Shakespeare's advice should be heeded as well, at least on occasion: "Have more than thou showest; speak less than thou knowest." It is amazing to me that the apostle Paul asks his faithful helper Timothy (II Tim. 4:13) to "bring the books . . . and parchments"[56] when he comes to visit Paul in the dark, dank Roman dungeon by the Tiber River,

---

[56]"God-breathed" from the Apostle Paul, New Testament, II Timothy 4:13, circa 67-68 A.D., written from the cold and cruel Mamertine Prison, Rome

even though Paul knows his time is just about up, as he *awaits execution*, but still with a desire to read and learn.

It goes without saying: obey the laws, and respect your fellow citizens. This is absent on some college campuses and public squares today. From his little notebook he carried in his small briefcase was found one of General Robert E. Lee's most important maxims for young men: "Obedience to lawful authorities is the foundation of manly character." Another timeless truth, indeed, but not always practiced by today's culture warriors. Be not disposed much to frivolous pursuits, like video games, social networking, TV shows, or romance novels and the like; commit to "redeem the time" in your daily life. Leaders and great men and women have discipline—they don't lead serendipitous daily lives. Years ago I heard someone say that the difference between a great man and a good man is what he does with his free time. A well-balanced life with success is a serious thing, not a some-time thing. It takes character and guts, and when you make it, you'll know you had 'em, and you'll feel good, in your core.

I have an old high school friend that signed off on a message once with *"vive fortis,"* Latin for "to live boldly." I liked that, but I would qualify it thusly: *with love*, and *under the rule of law* (except maybe the speed laws), and *by Jesus' statutes*. Realize also the real theme or virtue of nearly all achievement is victory over oneself. Crises are going to come, more than once; hopefully your good heart habits will pull you through. Perhaps also my favorite maxim from Thomas "Stonewall" Jackson: "Never take counsel with your fears." It's usually a mistake to make decisions out of fear (and biblical; see Job 3, Jeremiah 42, and Proverbs). The earlier manhood verse (I Corinthians 16:14) stresses men are commanded to do *all things with love*, even when acting as real men, while manifesting strength and courage. A slightly different form: "Perfect love casteth out fear" (I John 4:18). Another sage warrior friend put it this way: "always move *toward* your fears." Great advice, indeed, and one of my own life verse favorites is "God does not give you the spirit of fear, but of power, love, and a sound mind" (2 Timothy 1:7).

It's a given that you never get something for nothing. For some of you, it may be giving up watching NFL football and playing "fantasy football," spending too much time social networking, or playing in the over-the-hill rec leagues (still unable to shake, as Mr. Springsteen says, your *glory days*). You young men in transition need to get back to work, and/or help get other men back to work and be productive—not just for income—but to get them off self-respect-insulting government handouts, and to help them regain dignity. This modern-day availability of free stuff and govern-ment hand-outs is in opposition to the better policy of the old days, when the poor were offered a hand-up instead of a hand-out, and the emphasis was on faith, values and bedrock principles like the hard work, honesty, self-government, and self-control.[57] Speaking of self-control, found in Robert E. Lee's papers after his death was this: "A man cannot expect to control his children if he can't control himself." Back to work, confidence, and dignity; people are not happy unless they're productive. I hope you get to the point where you recognize that circumstances don't determine your potential, but it's your potential that God can use *to change* your circumstances. There are many inspirational American stories illustrating this. One of my favorite newer ones is that of preacher, speaker, writer, and former *trick* baby Ron Archer, whose mother was a prostitute. Check him out online or in person for an inspirational blessing.

Words, language and thoughts are so important. I'm almost embarrassed I even have to mention this, but part of my target audience here are millennials and slightly younger adults. The importance and power of proper and effective words, speech and language in this group is diminishing. It's unfortunate that many in today's selfie/smartphone culture don't seem to always value or realize their importance. Many teenagers and young adults are unfortunately not really literate, and are not able to harness and apply the power of language. Way back in high school I heard the famous defense attorney F. Lee Bailey (still going at 86), when asked for just one solid tip for success by an older student, said "Get a strong command of the English language." That surprised me, and was a tough task, but I never forgot it. Even the Bible talks repeatedly, over and over (especially in Proverbs), about the importance of speech, words, and language . . . and gets downright serious about it: "Death and life are in the power of the tongue" (Prov. 18:21). Also realize that as you work hard, live life, observe and interact with others, the use of profanity is usually evidence of a low self-image, and a non-Spirit-filled life. As one of my heroes Teddy Roosevelt said: "Profanity is the parlance of the fool. Why curse when there is such a magnificent language with which to discourse?"[58] Your words, thought and mental life are so important. Thoughts are things, very important things. A truth: "As a man thinketh in his heart, so is he." Powerful stuff, indeed; guard your heart. Focus on good, honorable, holy, truthful things . . . stay away from pornography, drugs and alcohol, foul language, and sexual immorality and remember: you are who you associate with. Hang around people who can lift you up, teach you, mentor you, and help you. As touched on earlier, all young people need inspirational

---

[57] For more on this, consider Howard Husock's *Who Killed Civil Society? The Rise of Big Government and Decline of Bourgeois Norms*, Encounter Books, 2019.

[58] Maurice Fulton, ed., *Roosevelt's Writings*, Macmillan, New York, 1922, p. x

models-or heroes-from history (including the Bible). Many of these popular American and biblical heroes are sadly ignored or attacked today by the secular left. My just mentioned personal model and strong Christian President Theodore Roosevelt was a thorn in the side to such leftists in the early 20th century. This was particularly true with eugenicist and Planned Parenthood founder Margaret Sanger, who was blinded by evolutionary pseudoscience. She claimed Roosevelt was "a disgraceful blight on upon any modern scientific nation's intent to advance."[59] Nevertheless "Teddy" Roosevelt recognized the importance of remembering our history and promoting our heroes:

> He wanted all men to be able to readily ascertain the chief ingredients of greatness and valor. He wanted to illustrate how thrift, industry, obedience to law, fulfillment of duty, and intellectual cultivation are essential qualities in the makeup of any successful people. He wanted to provide others with heroes such as heretofore faithfully guided him through the many shallows and shoals of his own life and career. He wanted all Americans to benefit as he had from the giants of valor.[60]

Part of the young man plan involves being a good father. My own father grew up in the great depression back in the coal mining country of the West Virginia hills, among the poorest of the poor. Somehow he survived and became a good three-sport athlete, a good Marine, a decent minor league left-handed professional baseball pitcher (until injury), national left-handed amateur golf champion after taking up the game in his mid-twenties, and successful in business. A memorable old novel opening line that always stuck with me, that I could relate to: "In my younger and more vulnerable years my father gave me some advice that I've been turning over in my mind ever since," from F. Scott Fitzgerald, in *The Great Gatsby* (1925). The most memorable occasion where this happened to me was early one morning when I was preparing to embark on a ten-hour drive to Illinois, by myself, heading off to college for the first time as an 18-yr. old freshman from the hills of West Virginia. Back then dad wasn't always the biggest talker; he was good at action. My stuff was packed and I was out

[59] Margaret Sanger, *The Woman Rebel*, June, 1914

[60] Theodore Roosevelt and Henry Cabot Lodge, *Hero Tales From American History*, Scribner's, New York, 1926, p. xxiii

in the garage, it was just starting to get light, things were a bit awkward as I prepared to leave, and some sort of goodbye was in order. This was one tough dude, but with a decent heart. He finally spoke, kind of low-key:

> "Son, I'll miss you . . . just do your best—work your tail off—
> but don't run your mouth or let the hard work show . . . and,
> I don't care who you know, where you are, or how successful
> you are, there's a thief everywhere you go, so watch your stuff.
> And no matter how things unfold, or how much money you
> ultimately make, always remember: you can count your true
> friends on one hand."

I must admit I was somewhat taken aback by the poignant moment, and a bit surprised by his carefully chosen words, but as the years went by I increasingly appreciated his sage advice. Dad was a good man then, and I knew he loved me, but he never told me—never *said* it—not that it was a huge deal at the time. That all changed some 32 years later in his mid-seventies, when he finally understood the real deal and accepted Jesus Christ as his Lord and Savior, and *God gave him a new heart* (what Jesus was trying to explain to Nicodemus, perhaps the greatest teacher in all of Israel, in the third chapter of John). It was truly amazing—a virtual *metamorphosis*—nothing else could have done it.

I recently chatted with a friend after he questioned me about the best evidences for recent creation, the Flood of Noah, the Ice Age, and Christianity. He knew I had been involved in those issues for years, and all of a sudden he was *hungry* for the Truth. After some quiet reflection, I slowly explained that I can talk at length about the manuscript and historical evidence for Christianity; the amazing reality of fulfilled biblical prophecy; the plethora of scientific evidence for recent creation and a cataclysmic worldwide Flood; and the evidence for *just one* major Ice Age, where such an event would have lasted hundreds of years, and required both cold *and hot,* which *only* a worldwide *mabbul* (Hebrew: powerful, all-encompassing flood) could provide. I added in this low-key discussion that creation scientists could thus explain the subsequent tremendous volume of locked-up water then in glaciers and ice sheets, and the subsequent lowering of worldwide sea levels, with the formation of continent-connecting land bridges, utilized by a divinely-forced dispersion of humans (the Confusion of Tongues at Babel), which led to not only different languages, but also the different skin colors and facial features, due to environmental differences acting on small and isolated gene pools. In other words, I explained that young-earth creation scientists, and myself to a lesser degree, could joyfully talk about historical, scientific, and other evidences supporting Christianity and the Genesis historical narrative *as written* (real history), including the lack of transitional fossil forms in the rock layers and new unfossilized soft tissue finds in dinosaur bones. I would not absolutely describe all of this as *proof* of the claims of Christianity, when teaching, answering a question, or witnessing for Jesus but could indeed share this as **mountains of strong evidence**. Nothing even remotely comparable is found in support of other world religions. As mentioned earlier, Christianity (and its forerunner, Judaism) is the only one *set* in history.

Here is **the main point,** though, for that somewhat tedious discourse: the real, best, most powerful beyond-a-shadow-of-a-doubt evidence for the Christian Gospel is ***changed human lives . . . like my dad's . . .* by God's Holy Spirit**. Nothing else could have done that. To see this tough, gritty, competitive, set-in-his-ways, successful and confident former athlete, United States Marine and businessman drastically change in his mid-seventies, and then *unashamedly* manifest and talk about *grace and love* was simply remarkable. It's not just *my* father; it is millions of other people throughout the centuries.[61] This reality of human transformation and the spirit of love—*getting a new heart*—was even recognized by Julian the Apostate, the last Roman Emperor who tried to hold to the old false gods and even implement New Age-type practices, in a desperate attempt to slow the transformation of the Roman Empire by the Gospel of Jesus Christ. It didn't work; the Christian juggernaut kept on growing. On his deathbed, after the battle of Samarra, he said this:

> You have won, oh Galilean. He [Jesus] has conquered Rome with love. His followers care more for their enemies, than we care for our friends.

Better late than never, as far as my dad goes; we had some great moments and discussions together, in his final years. He freely told me he loved me, on multiple occasions, possibly a little too much (trying to make up for lost time). What a heart change. Oh, I almost neglected to share another piece of his advice from that long-ago morning, perhaps pointing to his non-snowflake, no-safe-places early life (and well before he

---

[61] For a particularly moving and poignant American example, check Louis Zamperini in the **Glossary** at the end of this book, and his inspirational and riveting biography, *Unbroken*, by Laura Hillenbrand.

received Jesus), which surprised even me a bit: "Son, I don't care who you are, where you are, or how strong or tough you are, eventually everybody runs into somebody who can whip 'em!" Wow . . . all I could manage was a slight nod. This nugget, of course, while hinting at what some might call *toxic* masculinity, was virtually irrelevant to me and was simply filed away . . . as I fancied myself a lover, not a fighter.

Back to the importance of fatherhood- whether you're nineteen, a millennial, a baby boomer or even older—it is so important, and kids do better with fathers. In today's cultural confusion and leftist attack on manhood and fathers, many young people are understandably unsure or confused about the family, parenthood and what's right. God has called fathers to serious responsibility with leadership of the family. Even 4,000 years ago (Genesis 18:19), as God is referring to Abraham, we get the first mention relative to the father having the primary responsibility of *teaching* the children: "For I know him, that he will command his children and his household after him, and they shall keep the way of the Lord." In especially these modern times, many have dropped the ball, and real manhood and male leadership are trending in the wrong direction from political correctness, leftist academics, the sexual revolutionaries, No Fault divorce, and more. Men need to step up, and *fathers* need to be the primary source of solid answers to their kid's important questions. No-Fault divorce has destroyed thousands of families with countless children casualties, and it is men—the fathers—who are behind the eight ball. In his recent outstanding book, political scientist Stephen Baskerville points out how divorce stands today as the proudest celebration of feminine power. He adds: "Contrary to popular belief, the overwhelming majority of divorces— and virtually *all* involving children—are filed by women. Few involve specific grounds, such as desertion, adultery, or violence. The most frequent reasons given are nebulous and subjective: "growing apart," "not feeling loved or appreciated."[62]

Every father needs to remember that eventually his son, and in certain areas his daughter, will likely follow his example instead of his advice. Put another way, children will walk the way *you* walk, not the way you point. Always do the right thing; *walk the talk yourself.* In Matthew 3:17, Mark 1:11 and Luke 3:22 God says: "Thou art my beloved Son, in whom I am well pleased." Significantly, in all three cases, it is the *first use* of the key word *love* in each gospel. Although I am not a Bible scholar, I do know that when something is repeated in the Bible, and also when it *is first* mentioned or

---

[62] Baskerville, Ibid, p. 59

used, it is *extry* important, as Everette would say.[63] Significantly, the first use of the word "love" in the Old Testament is in Genesis 22:2, referring to the love of a father (Abraham) for his son (Isaac). Consequently God here is emphasizing that His love for His Son is the *very definition* of love. I realize this may not be popular with women's studies academics, psych doctors, and the transgender advocates, but it is clear and it is important. In John 17:24 it says that God the Father loved the Son before the creation of the world. In John's Gospel love is mentioned more times than in any other Bible book, the all-important first occurrence being "God so loved the world that He gave His only begotten Son" (John 3:16), that people might be saved.

Are you starting to sense that the Father-son relationship is important? The same applies to daughters, but naturally in a different way. A mother cannot give a father's love to a daughter (and a man cannot give a *mother's* love to a daughter). It is quite interesting in the Gospel of John that Jesus does and speaks *as taught by his Father.* In fact, in talking to his Jewish audience (John 8: 38) He says "I speak that which I have seen with my Father: and ye do that which ye have seen with your father." Remember, you millennials and dads, those young eyes and ears are always watching and listening *to you.*

---

[63] My 84 year-old friend Everette pronounces extra as *extry*, just like my favorite uncle from West Virginia did. He lives near me out in the country, works a 2-acre garden by himself, butchers hogs and deer (by himself) from October through late winter, makes the best pork sausage east of the Mississippi, can fix most anything, and is a dead shot and still proves this regularly on deer, rabbits, and varmints. He always subtly and quietly lets me know who has the strongest grip when we shake hands, usually accompanied by a brief steady stare. He has been married to the same woman for over 60 years, has great grown kids and grand-kids, and just recently gave up noodlin'. Noodlin' is slowly wading and feeling your way, with your arm underwater, along the stream bank, and finesse hand-fishing for those big-big lunker snoozin' catfish with your bare hands, and jerkin' those big whiskered slabs to the surface. He has significantly never grabbed a big snapper's mouth by mistake. He also makes the best slaw ever from his own cabbage (he won't share the recipe), and sells it to his friends for a moderate price, along with various vegetables and fruit (oh, those peaches!). Everette is a strong Christian, never uses profane language, raises Weimaraner dogs, and has shared some good country and outdoor wisdom with me . . . my kind of guy. He watches the news, and when chatting in the garden, I can sense he's not fond of gender fluidity, college safe places, political correctness, or socialism . . . and one can sense kind of a laid-back, subliminal hint of *toxic* masculinity. Everette is a real American man. If the bad guys invade, I would feel pretty good about sharing a foxhole with him. We could use a few more like him.

In touching on parenthood, thoughts, and heart habits, allow a brief return to the misunderstood cultural and personal focus on self, *self*-esteem, and *selfie* culture. The modern-day intense emphasis on these things is partly based on the false evolutionary ideas of psychoanalyst Sigmund Freud, and then other psych docs right on up to the *operant conditioning* and *self-esteem* theories of American psychiatrist B. F. Skinner. The foundational principle in Christianity of human beings with eternal souls being God's *image bearers* is denied by evolutionary psychologists, because of belief that man is just an accident of random cosmic happenstance, somehow evolving from an imaginary primeval soup. Many of their secular self-esteem-building remedies for human behavioral problems are unsuccessful long-term and are in opposition to Biblical principles. Attempting to boost a child's self-image, or even that of a confused young adult—when *undeserved*— can absolutely backfire. The Christian view of *others*-focus is actually opposite *to* the philosophy and worldview of most of the evolution-based movers and shakers of the late 1800s and early twentieth century. This includes Herbert Spencer, Friedrich Nietzsche, Francis Galton, Havelock Ellis, Margaret Sanger and many others (see Thomas Malthus section in chapter 3). They actually *promoted* selfishness, as well as neglect and *elimination* of the poor, the uneducated, the mentally or physically-compromised, and the chronically ill-not to mention certain *races*- as part of the desired effort to speed up the evolutionary process, and mocked the idea of *made in the image of God*. Influential Russian-American writer Ayn Rand (1905-1982) was strongly influenced by these intellectuals and atheistic culture warriors. Along with her two most famous books, *The Fountainhead* and *Atlas Shrugged*, she wrote a shorter one titled *The Virtue of Selfishness* in espousing her atheistic Objectivism philosophy and worldview.

I'll never forget seeing Ms. Rand on TV in a Phil Donahue interview, with the usual live studio audience, shortly before she passed away. I was in my twenties, and had read her three books just mentioned. At the very end of the show, Mr. Donahue asked her (I'm going from memory—will paraphrase), "So when you die, and you're buried in the ground, do you *really believe* that's it, and you essentially just rot away, and that's *the end* of your existence (no life after death)?" You could have heard a pin drop in the studio audience, and I was even on the edge of my seat at home. Ms. Rand was a bit stunned by the interviewer's closing, penetrating and poignant question . . . and she hesitated . . . a moment of reckoning. And then, almost reluctantly from the depths of her long-embraced, no-hope worldview, she said . . . "yes." It was then an uncomfortable, quiet moment on that TV set. The same applied to many in the viewing audience, including me, and I wasn't even a Christian at the time. But it got me thinking, long-term; obviously I never forgot it.

I hesitate to apply the overused phrase *diametrically opposed*, but these two discussed worldviews (Creation vs. Evolution)—relative to *self-image*—are indeed 180 degrees apart:

1) Made *in God's image*; called to be God's "image-bearers"; **others**-focused; **all** have extreme value.
2) Evolved *from primeval slime*; no God; no image importance-just another randomly evolved animal; *survival of the fittest*; **self**-focused; **many** human beings are simply *devoid* of value.

A good self-image has to be *earned* to be legitimate. Understanding *Who* made you locks it in.

PART II

SCIENCE AND THE CULTURE WAR

12

THE POLITICIZATION OF CLIMATE

THERE HAS BEEN great progress by the United States, the world's economic and industrial powerhouse, in cutting back on pollution and limiting the so-called *carbon footprint*. Other countries continue to struggle with this, or largely ignore it, especially countries that have an atheistic, Marxist, or socialist foundational worldview and history, or those nations without a Judeo-Christian origin or foundation, particularly China, India, Russia, Pakistan, Bangladesh, Egypt, Iran, Indonesia, several African countries, and even Japan. Nearly all of the 12-15 most polluted cities in the world are in India, where Hindu is the dominant religion. Recall my earlier claim that culture is religion externalized. In these larger and more populated countries as well as others (like United Arab Emirates, Bahrain, Mongolia, Qatar, Afghanistan, and Nepal), unhealthy and compromised water sources, air pollution, and lack of reasonable environmental protections are widespread, and they have significantly more serious pollution problems than the United States, Canada, and western Europe. Clean water is a huge problem in parts of Africa and southeast Asia. The *Climate Change* and environmental activist crowd, for the most part, **does not care** about these egregious international polluters and ecosystem assaulters-often *not one iota*. You can try to talk about it and show them the real data, you can share sickening pollution photos from these other countries, they can even go and view the pollution themselves, or you can show them the cholera and gastro-intestinal disease numbers, and they **don't care** – and it *does* not divert or sway their frantic and specious arguments against the industrial western nations, particularly the United States, one little bit.

It is now apparent that it's all about a political agenda, and ***power.*** Real facts and truth are not just ignored—they are *irrelevant*—and every crazy prediction by these alarmist climate prognosticators on the far left has proven false. This includes Al Gore's laughable climate prophecies, which followed Stanford biologist and Darwinian evolutionist Paul Ehrlich's ridiculous predictions 25 years before (he's still alive; be on guard). Even this past winter and spring many climate scientists were on the verge of a conniption fit, as there was irrefutable confirmation of glacier *growth*, particularly with Greenland's well-known Jakobshavn Glacier, previously considered a very rapidly-shrinking glacier. I'm sure they're scrambling and spinnin' things hard, though, to somehow morph this latest failed prediction into newfound and near-in-the-future, man-made climate change-disaster support. Consequently, it is now difficult, if not impossible, to have a reasoned, fact-based, civil discussion and debate on potential man-made influences on global climate and weather with the agenda-driven political left and weather alarmists.

This unfortunate reality applies to nearly all of the controversial issues bubbling up in this culture war (including gender dysphoria and the assault on men). If one doesn't accept and agree with the feminist, liberal, socialist, and now obvious Democrat Party line—*the facts be damned*—it is *hate speech*, and *racism*-yes, even on climate change. If you dare to even diplomatically challenge the claimed reality of significant man-made climate change you are now *racist*! There *is* climate change—there always has been—related to fluctuations in sun spots and solar wind, volcanism, and other factors, but the volatile disagreement is on *man-made* climate change. As mentioned, some leftist academics now even claim you're racist if you don't believe in man-made climate change! Furthermore, even if you sincerely claim you're *not* racist, you're a racist! A George Washington University professor recently came out with such outrageous claims and more in his book. Many decades ago my own father was friends with a fine man who was president of George Washington University; the school had then not quite fallen totally into the far-left group-think abyss. Just yesterday I read about a petition signed by George Washington University students protesting the white human walking indicator light at intersection crossings; it's just *too white supremacist-like* and racist. Looking at the metamorphosis of American universities through the years is simply stunning, as is the number of parents who pay exorbitant fees for secular, unscientific, and anti-American indoctrination. Parents: please do your homework relative to *where* you send or allow your kids to go to college . . . or even if college is the best option for them.

Back to **power**, or what my military friends call command and control. The term *climate change* is the bastard child of the former term *global warming*. The left had to come up with a more acceptable name because in the previous 15-25 years, there were a bit too many harsh, cold winters. Also, the term *climate change* is more comprehensive, with more serious far-reaching overtones, and is perhaps even more majestic-sounding, with the strong potential of additional government probing and extended tentacles into more areas of our lives . . . and thus, more command and control. A big part of the left's alarmist/fear mongering strategy is to sneak in socialism through a hidden back door. The climate change issue and agenda, like so many other new ones the last fifty years, has been unjustifiably forced on the American citizens by a small but effective minority, led by-once again-liberal academic professors and administrators, along with left-leaning mainstream media elites, the high priests of secular modernity. Hollywood is right there with them. Make no mistake about this issue: *Climate Change* gives the secular left a big tent to include-and, heaven forbid-implement, nearly every policy preference, radical change, or agenda proposal they have. This is **all about power,** and going after the capitalistic free-market system, and relieving those of us in *fly-over* country (the area between the liberal east and west coasts-full of people who farm, ranch, get their fingernails dirty, go to church, hunt, etc.) of rights, choices, and the little power we still have. China and Russia are most assuredly just smirking and laughing at our cultural discombobulation, and hoping some leftist climate change extremist gets elected President over Donald Trump, who is perhaps their worst nightmare.

The current catchphrase, or leftist talking point, is *"existential threat."* In other words, the contention is that dangerous man-made climate change is real, non-debatable, and is an *existential* threat (right now, baby!). What a joke; we do **not** have, or face, a climatological existential threat. I remember clearly earlier last summer (2019) when *every* Democrat presidential candidate *suddenly*-in the course of *one day*- used the term "existential threat" in talking about climate change. This was laughable, as it was so obviously a premeditated Democrat talking point to be universally stressed and repeated, because 1) the term was rarely, if ever, heard in the previous six months, and 2) suddenly-*in one day*-they were *all* using it, in virtually the *same sentence structure*. I just sat there for quite a while, scanning the news channels and Internet, chuckling and shaking my head. It happens all the time. Who puts out these Democrat directives and talking points?I'd bet money 90% of 'em never even used the word *existential* the previous year. You've got to give them credit, though— they recognize the importance of language—and are historically good at controlling the *terms* of the debate.

So, *existential threat* . . . it's such a moving, powerful, alarming and almost apocalyptic term, definitely worthy of a Hollywood thriller. The message: We are now in the battle of the ages—with our very lives and civilization itself at stake—against the most stalwart of foes . . . **climate change**, i.e., *the weather!* Nearly a century ago, American journalist H. L. Mencken opined: "The urge to save humanity is almost always only a false-face for the urge to rule it." Sadly, a significant percentage of our propagandized kids and young adults doubt the earth can even survive much longer, with unprecedented environmental disaster and massive worldwide extinctions, and the ultimate annihilation of human beings without massive policy changes, because of imminent catastrophic climate change. Part of the premeditated plan is to scare our impressionable children and students half to death. It is working;

our youth, especially in urban areas, are really afraid, as they're indoctrinated that multiple climate catastrophes are just over the horizon. They are taught that the earth is so delicate and fragile, while in reality it has been demonstrated over and over that our blue *water planet* is extremely tough, resilient, and self-cleansing. Many of these misinformed and manipulated kids, especially in Europe and the U.S., are so distraught they're receiving *eco-soothing* or *eco-friendly* counseling, while some are being treated with anti-anxiety drugs, and others have vowed to never have children (after all, who would want to bring a child into this quickly-dying world?!).

Even Democrat presidential candidate Andrew Yang recently said that people should be evacuated to higher ground (away from coastal areas) as soon as possible because of climate change (is he related to Al Gore?!). You just can't make this stuff up . . . and, where are the parents? Earth-shaking climate and environmental fear fits right in there very well with hate, homophobia, white privilege, white supremacy, white racism, white nationalism, victimology, identity and sexual politics, gender confusion, and all the rest. It does get weary, does it not? Emergency alert: *that's part of the plan*. Don't get lethargic and tune out, folks; get the real facts and get involved, and do not "get weary in well-doing." Stand for real science, the proven cyclical nature of weather and climate, real facts, and common sense.

The idea that man can control and direct, or change, the earth's weather and climate is borderline ridiculous to those of us who have spent a good chunk of our lives studying *the resilient earth*, analyzing it, mining it, extracting and utilizing its hydrocarbons and fossil fuels, hunting its animals, fishing, interacting with nature, ranching and farming it, and getting fingernails dirty and hands calloused working the land. At the same time we are blessed and astounded by observing the miracles of Creation and exquisite biological design, and all of the beneficial interactions, mutual dependencies and symbiotic relationships found in the natural world. Human beings that never experience this and grow up sequestered from nature in the cities often have a different worldview, especially if not part of a faith-based group, and are educated and indoctrinated more exclusively on the atheistic dogma of Darwinian evolution. They become more mesmerized by man-made things, instead of the miracles of creation all around them in the natural world, which are less obvious in large urban areas and the concrete jungle. As mentioned once already, the apostle Paul states that everyone is without excuse for non-belief in the Creator-God, because of "the invisible things of him from the creation of the world are clearly seen, being understood by the things that are made, even his eternal power and Godhead" (Romans 1:20).

Back to man, the weather, and the physical world; have environmental abuses and mistakes occurred? Absolutely, especially when there is ignorance of, or a drift away from the biblical call to be good stewards, in the primeval commission to mankind (Genesis 1: 28). In modern times, this was especially apparent in the late 1800s and early 1900s, as many western business magnates and industrialists, like Andrew Carnegie, John D. Rockefeller, Cecil Rhodes, and robber baron Leland Stanford (namesake of Stanford University), just to name a few, embraced and even financially supported racist social Darwinism and "survival of the fittest." History shows they abused their workers, and they abused and polluted the environment; they thought they *were justified* in doing so. The big chemical companies were guilty as well; this was very apparent in the 1900s where I grew up in the Kanawha valley around Charleston, West Virginia, with FMC, DuPont, Union Carbide, and other big chemical companies (they've cleaned their act up the last three decades). We're talking serious air and water pollution, but there were jobs and more jobs, and money to be made. As mentioned elsewhere in this book, this attitude and worldview was (and is) evolution-based, and it was felt by these people that what applies to biology applies to society. This fortunately fell out of favor with Nazism, Adolph Hitler, the tragedy of the Holocaust and 75 million military and civilian dead in WWII. But a major point: when the bulk of the population is *urban*, and has no real connection to the land or ecosystem, or an understanding of where important foundational things *come from* (food, building materials, fossil fuels, steel, aluminum, etc.), nor an appreciation of the miracles in the natural world, they are more easily swayed and manipulated.

An inspirational hero of mine is former conservationist, scientist, forester and the father of American wildlife management, Aldo Leopold. Mr. Leopold, who many conservationists, ecologists, wildlife lovers and hunters consider a national treasure, wrote *A Sand County Almanac* and later its follow-up essay, simply called *Land Ethic*, which was published in 1949 shortly after his death while fighting a prairie fire in Wisconsin. *Land Ethic* was a call for moral responsibility to the natural world, while studying it and being good stewards, caring for it *and* people, and strengthening the relationships between them. Mr. Leopold would likely be stunned and disappointed by the disconnect between our youth and the land and natural world today, and the grim alarmist and false-science indoctrination. My earlier claim of ripeness for manipulation is especially true if they are ungrounded, with no Judeo-Christian upbringing, or faith-based perspective. My friend Pat was recently talking about an experience with underprivileged kids from the city that he and a friend took for a ride out in the country. One little boy about nine years old pointed to an obvious cow out in the green pasture and asked "What is that?!" Pat was a bit stunned, and simply replied, "Well . . . that's a cow." The curious boy then asked "What's it for?"

Just last week (autumn, 2019) a new young hero of the political left, 16-year old Greta Thunberg of Sweden, who has been diagnosed with Asperger's

Syndrome, testified before the United States House of Representatives on imminent catastrophic climate change. (As a sidebar note, for decades Sweden has been considered by many to be one of the most secular countries in the world.) My initial reaction to Ms. Thunberg was *disbelief* when I heard about this young lady's testimony. How and why did this happen? Who orchestrated this? What are the young Swede's qualifications, training, and earned academic degrees? Is she simply precocious in knowledge of ecosystem dynamics, the physics of sun spot and solar wind fluctuations and their effects on the Earth's cyclical, but expected, mild vicissitudes of temperature? Does she realize that good ol' water vapor is the #1 greenhouse gas? Most on the political left along with the loud young climate-activists do not.

Don't get mad; down here in Carolina we just didn't have much scoop on Ms. Thunberg—so I admit my ignorance—I'm just asking. Was she a child prodigy, perhaps, on the amazing but complicated atmospheric hydrological system of the earth, or in the discipline of meteorology, or the esoterica of geophysics? How did she become such an expert? It was revealed that she traveled across the big pond by huge sailboat for supposed carbon footprint minimizing, with no mention—of course— that the boat probably had to be returned, sailing all the way *back across* the Atlantic Ocean, with its possible use of fuel and oil, if the motor is used. Those solar panels on her boat, the Malizia II racing yacht, don't produce a lot of energy to power the underwater turbines in cloudy, windy, inclement weather (common in the early winter north Atlantic). Perhaps the boat will be sold in America; the crew ironically flew back on jet aircraft. Wow, talk about carbon emissions! Little Greta, presently speaking in Canada (Oct., 2019), still doesn't know how she's getting back home across the pond.

The large climate alarmist youth movement is now known as *Extinction Rebellion* (or just XR), and believes the earth is flat-out doomed without a total metamorphosis of the global economy and a cessation of fossil fuel use. This ridiculous ungrounded scaremongering is simply stunning. As soon as she arrived at the marina in New York, Ms. Thunberg spoke on the supposed climate crisis, as *"the biggest crisis humanity has ever faced."* Wow, just an absolutely amazing statement, *if* she is indeed an expert, or at least incredibly knowledgeable and informed, about other major crises like the Crusades and the later assault on Europe by Islam; the Black Death (bubonic plague) of the 14th century, with over 20 million deaths in Europe (1347-1352), after prior millions in Asia in the early 1340's; and later the tragedy of WWI, with 40 million casualties ("the war to end all wars," according to Woodrow Wilson and H. G. Wells). Then there's the 1918 Spanish Flu epidemic which killed 20 million to 50 million victims (including 675,000 Americans). I wonder if she is also schooled on the unimaginable tragedy of Japanese Imperialism and Hitler's applied Darwinism of WWII— with 75 million of war and civilian casualties worldwide—and communist Russia and the United States coming to the brink of unthinkable catastrophic nuclear war just several decades ago—just to name a few.

This entire situation would be comical if not for the millions of propagandized, scared kids who think their life and that of the earth is virtually over. The dearth of knowledge concerning real science, real history, and the resilient earth is simply profound. CNN chief climate correspondent Bill Weir glowingly described her as the "old soul of the climate crisis," and with her congressional testimony, coming off "as the *oldest soul* on Capitol Hill." Some of these climate change people are so off-base, misguided, misinformed, and even borderline crazy . . . juxtaposed with their secular coaches and manipulators behind the scenes that are "wise as serpents." Politicians and business leaders can't say that, but I'm saying it. It is preposterous on so many fronts, and it is just a power grab. Rookie U. S. Representative and socialist New York Democrat Alexandria Ocasio-Cortez, the new female Congresswoman representing one of the filthiest, most garbage-laden districts in America (the eastern part of the Bronx and part of north-central Queens, in N.Y. City), has been feted by the mainstream media and far left from the get-go. Chiming in repeatedly on make-believe man-made catastrophic climate change, she uttered this ungrounded statement in early 2019, with several lies in one sentence: "There's a scientific consensus that the lives of children are going to be very difficult, and it does lead, I think, young people to have a legitimate question: Is it OK to still have children?" These leftist Democrat "existential threat" politicians in America couldn't care less about the environment, and the people's welfare. Just look at the major cities *they've run for years*: Seattle, San Francisco, Oakland, Los Angeles, Chicago, Detroit, St. Louis, New Orleans, Baltimore, Washington, D.C., New York and now even to a lesser degree, Boston. Look at their filthy tent cities everywhere, the drug paraphernalia and used needles scattered about, the trash, the vermin, the passed out humans and feces on the sidewalks and parking lots, the homeless, the beggars, and the murder rate, not to mention the missing, disappearing tax dollars, and the non-cooperation with law enforcement. Public schools and education are collapsing in some of these cities, and of course the kids can't make it if they can't speak English. It is an unbelievable national disgrace and embarrassment, and now double-dangerous with the potential epidemic of serious disease (especially in San Francisco and New York).

With the absence of a faith-based worldview, improper grounding, no Biblical glasses to put on to properly see the world, along with a disconnect from the land, these young people have no filter or grid to properly consider and evaluate the preposterous claims and doomsday environmental scare-tactics regularly thrown at them by the liberal left. They are so ea" manipulated. I think of them as young duped human pawns ready for the picking. As previously mentioned, the far-left—led by academic elites, the mainstream media, with psych doc and white lab coat support, and now Big Tech— have unfortunately been successful in suppressing belief in the Creator God, challenging Biblical authority, and promoting the idea that there is no transcendent purpose outside of and greater than ourselves. After all (according to them), we're just another accidentally-evolved animal. It takes such a resultant materialistic, no-hope, limited worldview for young people to seriously consider and believe such climate change drivel and *existential threat* nonsense. I had a sharp mentor many years ago who said if you don't stand for something, you'll fall for anything. You see this throughout our culture today. Be smart, and remember those timeless words from Proverbs: the *beginning of knowledge and wisdom* come from the fear of the Lord.[64]

---

[64] See the *Brief Mentions* chapter, for more on the *"Fear"* of the Lord

13

EVOLUTION: TRANSITIONAL FORMS IN THE FOSSIL RECORD? MISINFORMATION ALERT: THERE ARE NONE

I AM IN MY mid-sixties now-a seasoned citizen-but in my science classes all through school, from about 6th grade through college, it was stressed that the so-called fossil record was the main area of evidence for the "theory" of evolution, especially relative to biology and geology. I hesitate to use the word "theory" with evolution as it does not even come close to qualifying as a scientific theory. Evolution, or Darwinism, is an idea, a concept, a story, maybe metaphysics—some would say even a religious belief—but *not* a scientific theory. A scientific theory—by definition—is a formal explanation of some phenomenon, process or principle in the world around us based on empirical evidence. The evolution idea thus does not qualify. Many feel it does not even qualify as a hypothesis, while although the word *hypothesis* does relate to presuppositions and unverified scientific ideas or conjectures (like evolution), a legitimate hypothesis is based on facts, and cannot contradict the known laws of nature—like the never-disproved, no exceptions to, 1st and 2nd Laws of Thermodynamics—which the evolution idea egregiously does.

One needs to understand that if evolution is true, and there has been slow and gradual transformation of organisms into completely new and different types over the last 500 million years or so, with even the *higher* animals and human beings ultimately originating from-as Carl Sagan (of Cosmos fame) loved to say-"the primeval soup," then there **should be millions upon millions** of intermediate (or *transitional*) forms found in the fossil record. **There are none.** Even Charles Darwin himself was puzzled:

Why then is not every geological formation and every stratum full of such intermediate links? Geology assuredly *does not reveal* any such finely graduated organic chain; and this, perhaps, is the most obvious and serious objection which can be urged against the theory. The explanation lies, as I believe, in the extreme imperfection of the geological record.[65]

Darwin was confident, or at least hopeful, that eager and resourceful paleontologists and fossil sleuths would discover such intermediate forms in the decades ahead as the sedimentary rock layers revealed their secrets. Disappointment continued. In his "Life and Letters," Darwin admitted "**not one change of species into another is on record . . . we cannot prove a single species has changed into another**" (there should be gobs of 'em). Consequently, let's fast forward a bit-say 120 years-to get an update from Dr. David Raup, the former Curator of Geology at the well-known Field Museum of Natural History in Chicago, on this all-important issue:

Well, we are now about 120 years after Darwin (circa 1980) and the knowledge of therecord has been greatly expanded. We now have a quarter of a million fossil species but the situation hasn't changed much . . . we have *even fewer* examples of evolutionary transition than we had in Darwin's time . . . some of the classic cases had to be discarded . . . So Darwin's problem has not been alleviated in the last 120 years . . .[66]

For some readers this may be surprising; this is *not* what you were taught in biology class, paleontology textbooks, or natural history museums. Dr. Raup goes on:

Darwin's theory of natural selection has always been closely linked to evidence from fossils, and probably most people assume that fossils provide a very important part of the general

[65] Charles Darwin, 'On the imperfection of the geological record', Chapter X, *The Origin of Species*, J. M. Dent & Sons Ltd, London, 1971, pp. 292-293

[66] Dr. David M. Raup, 'Conflicts between Darwin and paleontology', *Field Museum of Natural History Bulletin*, vol. 50(1), Jan. 1979, p. 25

argument that is made in favor of Darwinian interpretations of the history of life. Unfortunately, this is *not* strictly true.[67]

Actually, it is **not true at all**. If it's not true, then why was I taught this in school, along with millions of other unsuspecting students? I always wondered about this, even when in high school, thinking how in the world could an intermediate form *even survive*? For example, when taught that the wing of a bird most likely evolved from the forelimb of a reptile, one justifiably wonders how such a creature could *even function* in such a hybrid, compromised condition (it couldn't, of course). Stephen Jay Gould, now deceased but former paleontologist and famous evolution spokesman of Harvard University, said this:

> The absence of fossil evidence for intermediary stages between major transitions in organic design, indeed *our inability, even in our imagination,* to construct functional intermediates in many cases, has been a persistent and nagging problem for gradualistic accounts of evolution.[68]

Indeed, a stunning and honest admission . . . but way too late, for millions of duped students and fragile teenage, faith-based worldviews. Gould also stated:

> All paleontologists know that the fossil record contains precious little in the way of intermediate forms; transitions between major groups are characteristically abrupt.[69]

Again, wow! These are significant admissions by a Harvard professor who was one of the leading spokespersons for evolutionary theory in the

---

[67] Raup, ibid, p. 22

[68] Stephen J. Gould, 'Is a new and general theory of evolution emerging?' *Paleobiology*, vol. 6(1), Jan. 1980, p. 127

[69] Stephen J. Gould, 'The return of hopeful monsters', *Natural History*, vol. LXXXVI (6), June-July 1977, p. 24

second half of the twentieth century. **Why is this not in the textbooks**?! In many people's minds, the main proof or evidence confirming evolution is the revealed fossil record in the rock layers around the world. The entire thing reminds me of another controversial issue years ago where Kay Cole James said: "You're entitled to your opinions but not your own set of facts" (Patrick Moynihan said that first-I think).

You may be thinking "not so fast!—the textbooks, the museums, the TV pundits, even the movies all talk about it—it's got to be true . . . and there just has to be fossil evidence supporting evolution!" Just hold your horses; let's consider what another highly respected and international-ly-known scientist had to say, British paleontologist Dr. Colin Patterson. Dr. Patterson was eminently qualified to comment on this subject, as he was the long-time Curator of the British Museum of Natural History in London, which houses more fossils than any museum in the world. I visited this world -famous museum years ago and recommend it, but make sure to put on your best propaganda- discernment glasses, as this edifice has been the sacred shrine for long ages evolutionary dogma for the last century and a half. Several years back, when asked why his new book on evolution contained **no** pictures or illustrations of transitional (intermediate) fossils, Dr. Patterson amazingly stated:

> I fully agree with your comments on the lack of direct illustration of evolutionary transitions in my book. If I knew of *any,* fossil or living, I would have certainly included them . . . Gould and the American Museum people are hard to contradict when they say there **are no transitional fossils** . . . You say that I should at least 'show a photo of the fossil from which each type of organism was derived.' I will lay it on the line—there is **not one** such fossil for which one could make a watertight argument.[70]

Sharp millennials, college students, and frustrated truth seekers—let me repeat: ***not one*** *such fossil . . . not one*—that you could "make a water-tight argument" for, this from a highly respected international expert. If conventional slow and gradual macroevolution over hundreds of millions

---

[70] This honest revelation is from a personal letter—written Apr. 10, 1979—from the highly respected Dr. Patterson to Luther Sunderland, as quoted in *Darwin's Enigma* by Luther D. Sunderland, Master Books, San Diego, CA, 1984, p. 89. (Newer editions, Master Books, Green Forest, AR)

of years is true, there should not just be gobs of 'em, but *many millions* of such intermediate forms. Over a billion fossils have now been found, but it is astonishing that **not one** clear-cut, for sure, undisputed intermediate fossil form has been found to support the worldview that has dominated academia, science, social policy, psychiatry, much government policy and so much more in the last century. Are you beginning to sense you've been deceived?

Even though I disagreed with Stephen Jay Gould's admitted atheistic and evolutionary beliefs, I respected him for honestly admitting **the absence of** transitional forms in the fossil record and strongly promoting to his peers the need for a new idea—a new paradigm. Even this mild compliment is a tough admission for me, as the famous Mr. Gould's marxist worldview denied and virtually mocked the biblical Creator-God's primary focus: *to create and redeem men and women in His image*. The Harvard professor's less than uplifting (putting it mildly) opposing view:

> Man . . . or even woman . . . as the crowning achievement of some grand cosmic plan? What moral conceit. We are but an afterthought. We are a little accidental twig.

Nevertheless, he and his like-minded colleague Niles Eldridge came up with the idea of "punctuated equilibria" decades ago, which basically claims that natural selection/evolution must have been marked by isolated episodes of very rapid and profound evolutionary change, between long periods of little or no change at all. Huh? Say again? The main idea of this odd about-face proposal is that evolutionary change happened *so fast*, there were no transitional fossils left. Some evolutionists reluctantly agree with this fallback argument, many do not. So, let's get this straight: evolution occurs so **slowly** in the **present** we can't see it, but occurred so **fast** in the **past** it didn't leave any evidence (and we can't see that either). Hmm-mm . . . I'm thinking, you can't make this stuff up . . . but . . . they have. Each one of you should look at the real scientific evidence, or the lack thereof, and then decide. I am very aware of the difficulty in overcoming years of indoctrination and propaganda in public school science classes and textbooks, natural history museums, movies, and on TV and national park signs. This is heady stuff. How you view your origins affects your entire worldview—and how you live your life. It is just amazing how entrenched but unsupported assumptions and presuppositions—like billions of years of time—affect one's interpretation of the evidence, is it not? (See, *The Problem with Assumptions*, Part III). Even the well-known

American husband-wife historian team of Will and Ariel Durant (not professing Christians), who both passed away in 1981, had this to say: "Most
history is guessing, and the rest is prejudice." Back to internationally-recognized fossil expert Dr. Gould one last time:

> The extreme rarity of transitional forms in the fossil record
> persists as the trade secret of paleontology. The evolutionary
> trees that adorn our textbooks have data only at the tips
> and nodes of their branches; the rest is inference, however
> reasonable, *not the evidence* of fossils . . . We fancy ourselves
> as the true students of life's history, yet to preserve our
> favored account of evolution by natural selection, we view
> our data *as so bad* that we **never see** the very process we
> profess to study.[71]

Shocking . . . again . . . *we never see*! Please read that again—from a former
internationally recognized evolution/long ages "expert"—we *never see* . .
. evolution by natural selection, or Darwin's famous "tree of life" . . . in the
fossil record. No evidence of transitional (or intermediate) fossil forms, no
real evidence anywhere of vertical or new organism/body-type evolution,
in the present—or in the rock strata—and additionally, no known mechanism of evolution's required addition of new genetic information to the
genome (the DNA software code) for such a process to even be feasible.
And yet, this evolution idea took hold and changed the world. In Charles
Aughton's book *Voyages That Changed the World*, in referring to the young
Charles Darwin's 5-year around the world sailing voyage as naturalist on
the *HMS Beagle* in the 1830's, he states: "In many senses it changed the
world more than any other in history." Relative to the *evolution by natural
selection* idea birthed by that long voyage, Dr. Michael Denton summarizes
the eventual effect of it, and Darwin's subsequent *Origins* book, as

> a new and revolutionary view of the living world which
> implied that all of the diversity of life on Earth had resulted
> from natural and random processes and not, as was previously
> believed, from the creative activity of God. The acceptance
> of this great claim and the consequent elimination of God

---

[71] Stephen J. Gould, 'Evolution's erratic pace', Natural History, vol. LXXXVI(5), May
1977, p. 14

from nature was to play a decisive role in **the secularization of western society**. The voyage on the *Beagle* was therefore a journey of awesome significance. Its object was to survey Patagonia; *its result was to shake the foundations of western thought.* [72]

Anthropologist and biology/medical writer Richard Grossinger put it this way:

Charles Darwin and Karl Marx (an enthusiastic supporter of Charles Darwin-*RJ*) changed human reality to such a degree that a gap formed between the landscape brought into being by their visions and the world they inherited.[73]

A *gap* formed?! A *gigantic canyon* would be a far more accurate term. Thomas Huxley, the most vocal and influential early promoter of evolution-known as "Darwin's bulldog"-unashamedly stated: "It is clear that the doctrine of evolution is directly antagonistic to that of creation . . . if consistently accepted, **it makes it *impossible* to believe the Bible**." No truer statement was ever uttered by an evolutionist. But, let's go slow, and reason together, as we examine one of evolution's largest elephants in the living room.

---

[72] Michael Denton, *Evolution: A Theory in Crisis*, Adler & Adler, Chevy Chase, Md. 1986, p. 17.

[73] Richard Grossinger, *Homeopathy The Great Riddle*, North Atlantic Books, Berkeley, Ca., 1998, p. 11.

14

CARBON DATING, *FAST AND FURIOUS*, AND HARVARD STORYTELLING. . . . "C'MON, MAN!"

LIFE AS WE know it requires many specific elements and conditions, but it is carbon-based. Carbon makes up less than 0.1 percent of the Earth's crust, yet it is the element most essential for life. The word *organic* implies living organisms, or formerly living ones, or carbon-based compounds. Diamonds, graphite, natural gas, petroleum and coal are also made of carbon, the latter three from carbon-based buried life forms-mostly plant matter-so their age can be somewhat tested with radiocarbon dating-often referred to as Carbon-14 dating (Carbon 12 is normal and stable). Being unstable, or radioactive, such buried (but formerly living) organic matter gives off at a certain rate its volatile, unstable Carbon 14 portion. The Carbon-14 portion makes up just a tiny percentage of an organism's carbon; the decay rate, or "half-life" of this carbon is 5,730 years. After this specific time period, half of any original amount of the Carbon 14 disappears. After another 5,730 years transpires (the second half-life), only a quarter remains. As this fast radio-decay continues, and after three half-lives only an eighth is left, and so on to the point where not a single Carbon 14 atom should exist in any carbon source older than about 240,000 years. If the Genesis historical narrative is true, with a wonderful worldwide mild subtropical climate and the likely ultra-rich biosphere in the antediluvian era, Carbon-14 concentrations in life forms would have been much lower than modern levels (distributed through *higher numbers of* plant and animal life forms than after the Flood). Remnant Carbon-14, though, would still be expected in small amounts in organic sedimentary discoveries—including dinosaur bones. Consequently if *any* Carbon-14 is found in ancient organic matter, it is proof that the source organism or former plant matter *cannot possibly* be millions of years old. It is now known that all coal seams—previously assumed (and taught) to be anywhere from 30 to 300 million years old— contain remaining radiocarbon at an average level *250 times greater* than the minimum detection level of the sensitive equipment. **Yet none should be found[74] if the coal is older than 200,000+ years**.

We are taught that diamonds formed 1-3 billion years ago, along with the supposed hundreds of millions of years for coal. Evolution advocates need these tremendous ages to justify their worldview, even though evolution *has never been observed in the present world or in the fossil record*, and goes against the most proved laws of science. Most of you have been indoctrinated that evolutionary change is demonstrated in the fossils found in the sedimentary rock layers around the world, but that is **absolutely 100% false**. Please understand that *Deep Time*, what I refer to as long ages—billions of years—is the #1 cornerstone requirement to even have a remote chance to explain the absolute miracle of design and life without God.

Let's set aside academic and scientific storytelling, and look at the facts. Legitimate certified radiocarbon laboratories have reported that diamonds, like coal, have significant remaining Carbon 14 amounts at many levels over the detection threshold. Petroleum (liquid hydrocarbons) and dinosaur bones brought up from the sedimentary rock strata also contain this unstable radiocarbon. This is incredibly strong evidence for **young** age—*thousands* of years—not millions or billions. The ramifications are profound. Coal has actually been made in the laboratory from various types of compressed organic matter (leaves, bark, limbs, organic debris and remains), under pressure with heat production, in mere *days to weeks*. Such conditions may well simulate massive mats or amounts of destroyed vegetation (and some animals) from a past catastrophic world-wide flood, with huge layers of mud and sediment involved and tremendous pressures and tectonic forces. In more recent but similar experiments, coal has even been formed quickly *without* pressure.

One can easily find very well-preserved leaf and insect fossils in coal seams; I have found them myself in multiple locations in my home state of West Virginia and out west in the Dakotas and Wyoming. This screams *rapid* burial, which would have been possible only in a rapid and prodigious sediment-laden, large-scale flood event, required to envelop and protect the plants and animals from deterioration, rotting, or being eaten. We are talking about, at least in some locations, almost unimaginable massive and thick layers of coal made from a huge biomass of dead plant material, transported quickly and buried before it decayed. In Wyoming, which long ago passed my home state of West Virginia as the leader in coal production, there are found at least six coal beds in the Powder River Basin exceeding 100 feet in thickness (over 200 ft. in some places), with some of these beds amazingly extending under the surface (below prime antelope country) for over 75 miles. Coal seams are found

---

[74] Dr. Larry Vardiman, *Radioisotopes and the Age of the Earth*: Results of a Young Earth Creationist Research Initiative, Master Books, Green Forest, AR, 2005.

around the world in intermittent, repeated overlying layers, separated by various sterile rock layers with no sign of soil. This irrefutably indicates largescale deposition and burial under water, with the rock layer sediment washing in between the layers of coal (formerly Flood-destroyed matted plant material).

These sedimentary rock layers cover and compose 75% of the earth's land surface, are undeniably laid down by water, and vary from a few feet *to many miles thick* (with billions of fossils), with the average thickness being 1.5 miles! They are full of dead things, and are even found underneath the ice and snow on top of the tallest mountain ranges in the world, containing estimated billions of fossilized *marine* organisms. Even famous Mt. Everest (and the other Himalayas) is mostly water-deposited sedimentary rock (later uplifted). Conversely, one of planet earth's other great peaks, mighty Mauna Kea in Hawaii, rises over 30,000 feet from the ocean bottom, but an unbelievable 56,000 feet from its deep underground mountain foundation, and is made entirely of solidified volcanic lava. Although of different composition, creation scientists believe both of these great mountains were formed during or soon after the worldwide Deluge of Noah's day. Evolutionists deny this—partially due to the current slow rate of volcano growth— but such slow and gradual *uniformitarian* processes seen today do not always apply to the past. The scientific and historical findings—discussed repeatedly in this book— unequivocally point to a past period, 4 to 5 thousand years ago, of *worldwide* catastrophic processes, best described as **fast and furious**. The skeptics and evolutionists also deny the reality that at the beginning of the Great Flood, "*all* the fountains of the great deep [were] broken up"(Genesis 7: 11). This would have caused not only earthquakes, volcanic eruptions, destructive super-tsunamis, turbidites, hypercanes and tectonic plate movement but massive inflows of hot molten lava at mega-increased rates over what we see today. As mentioned before, if you deny the global earth-reforming Flood, you misinterpret everything else. Massive Mauna Kea in Hawaii could have easily formed some 4,300 years ago under such volatile and violent conditions.

Of course, you probably didn't learn this in censored public school and college biology and geology textbooks. It doesn't fit the secular paradigm, so it is purposely kept out of the textbooks. Back to the *right conditions* for rapid coal and fossil formation, diamonds can even be formed very fast in the laboratory, as well as opals (just a few months). It is now very clear: the formation of fossils, oil or petroleum, natural gas, coal and diamonds doesn't take millions or billions of years—it *just takes the right conditions*—which a catastrophic world-wide flood would provide. What do evolution/billions of years advocates have to say about these radiocarbon finds? The answer: virtually nothing. It's an elephant in the living room. It is simply ignored.

The Bible spends a significant 4 chapters in Genesis discussing the Ark of Noah and a cataclysmic world-wide flood, and it is referred to or mentioned many times later in scripture, even in the New Testament. If true, it is one of the singular events in world history. There is *not a hint* of long ages anywhere in the Bible. The great Flood is referred to multiple times in the Bible besides the book of Genesis, especially in Job and in the New Testament-even by Jesus. Specifically in regard to those who ignored Noah's message and invite, Jesus himself said "For as in the days that were before the flood they were eating and drinking, marrying and giving in marriage, until the day that Noah entered the ark, And knew not until the flood came, and took them all away; so shall also the coming of the Son of man be" (Matthew 24: 38-39). Such an earth-reforming Deluge would have occurred some 4,300 years ago, based on real historical and scientific data as well as Old Testament genealogy studies, backtracking from Jesus Christ through David to Jacob, on back through Abraham, then Noah and finally the 1,800 or so antediluvian years back to Seth and Adam. This was first chronicled in detail by one of the greatest historians and Bible scholars in the 17th century, Archbishop James Ussher, in his classic *The Annals of the World*, where he painstakingly came up with the date of Creation as around 4004 B. C. Dr. James Ussher is not to be taken lightly. He entered Trinity College at age thirteen, received his Master's Degree at age eighteen, earned his Doctor of Divinity at age twenty-six, and was an expert in Semitic languages. He was the Church of Ireland Archbishop of Armagh, head ecclesiastical *Primate* of all Ireland between 1625 and 1656, and a meticulous historian and scholar who devoted his life to the defense of the Christian faith. He was buried in 1656 in Westminster Abbey (ironically, the same place as atheist Charles Darwin, over two centuries later).

What if Ussher's date is accurate and such a world-changing global Flood really did occur, some 1700-1800 years after the Creation date, some 4,350 years ago? What would you expect to find? A major part of the answer: extensive fossil fuels (coal, petroleum, natural gas) and unimaginable numbers of remnant dead organisms in water-deposited sedimentary rock layers—all around the world—most likely on all seven continents. Remember it doesn't take millions of years; it just takes the right conditions. Even Charles Darwin himself, in his writings and notes about the Galapagos Islands in 1835 from his famous voyage, in describing the mountains there that were obviously crowned with their easily-identifiable craters, and the boundaries of their young-looking lava streams still distinct, stated "we are led to believe that within a period, **geologically recent**, the unbroken ocean was here spread out." This calls to mind the Noahic Flood scripture just mentioned, "The *same* day were *all* the fountains of the great deep broken up, and the windows of heaven were opened" (Genesis 7:11).

For a look at remnant evidence of this, check a map of the geological fault lines, fissures, seafloor topography and specifically the mid-Atlantic ridge running the length of the Atlantic Ocean floor. The young English naturalist, in those same notes, in describing the strange giant tortoises there, ironically used the adjective *antediluvian*, which of course can mean extremely old-looking, but refers specifically to the time before the biblical flood.

Darwin's use of this adjective seems sardonic as well as ironic, as his evolution idea and *Origins* book (published in 1859) were the main catalysts for the increased eisegesis of biblical texts and accelerated disbelief in Genesis history and the Flood of Noah. This quickly led to the loss of biblical authority by the intellectual elites and many others of the late nineteenth century and beyond. The groundwork for this rapid acceptance of his idea was already laid down by Scotsman Charles Lyell (an encourager and mentor of Darwin) and his three-volume *Principles of Geology* decades earlier. One needs to understand that geology as a recognized separate scientific discipline is only a little over 200 years old. The young Charles Darwin devoured these books on his 5-year voyage. Various other influential amateur geologists agreed with Lyell and his long ages-uniformitarian views. This transformation of historical and geological thought and the concomitant erosion of biblical authority is all documented and explained impressively in *The Great Turning Point*, by Dr. Terry Mortenson.[75] Lyell and Darwin's long-ages/evolutionary ideas were strongly attacked and criticized by contemporaries and strong Christians James Clerk Maxwell and Michael Faraday, among others, unquestionably two of the most brilliant physicists and inventors not only of their day, but of all time. Maxwell, who was later greatly respected and admired-if not idolized-by Einstein himself, even wrote an impressive scientific paper attacking the *Origins* book and the impossibility of its evolutionary idea; his argument has never been refuted. I give Darwin credit for being a good naturalist, but using an old athletics cliché, he couldn't carry James Clerk Maxwell's jockstrap. Incidentally, the brilliant Maxwell had Psalm 111:2 inscribed above the door to the famous Cavendish Laboratories at Cambridge. Here is one of his inspirational life verses: "The works of the Lord are great, sought out of all them that have pleasure therein."

One of my favorite heroes, Michael Faraday, the humble self-educated experimental physics genius and inventor, defended the Creation account and Genesis history until his death. Faraday, who with little education somehow emerged from the London slums and helped change the world with his inventions and knowledge of physics, light, optics, magnetism, electricity, their interactions and so much more, refused offers of

---

[75] Terry Mortenson, *The Great Turning Point,* Master Books, Green Forest, AR, 2004

knighthood and honored burial at Westminster Abbey, claiming he was just a humble sinner *"trying to think God's thoughts after him."*

Despite his admission of the terrain appearing "geologically recent," and the use of the adjective *antediluvian* in describing the Galapagos Mountains and reptiles, Darwin went the way of his friend and mentor Charles Lyell and his ungrounded uniformitarian idea. He reasoned that if *slow and gradual* change over long ages applied to geology, it must also apply to biology. This view denies catastrophism, and any worldwide flood, despite a plethora of evidence. As the young naturalist spent many enjoyable hours hiking the rugged hills in the southern hemisphere (romanticized in the Stephen Maturin character, as part of the excellent 2003 movie *Master and Commander*), he wondered at how so many sea creatures came to be fossilized at many thousands of feet above sea level (how 'bout a big flood, Mr. Darwin?!). So what do we actually find? Nobody says it better than Ken Ham of the *Answers in Genesis* ministry (which now has the world-famous Creation Museum and Noah's Ark Exhibit in northern Kentucky): "All you find are billions of dead things, buried in rock layers, laid down by water all over the earth." And yes, even in Alaska, Greenland, Siberia and under the Antarctic ice (some 2 miles thick). Keep in mind that fossil formation today is *extremely rare* and difficult, and in recent decades even secular geologists have reluctantly admitted or adopted *neo-catastrophism*. Richard and Tina Kleiss offer a logical perspective:

> The very nature of the fossil record testifies to a worldwide water catastrophe in the past. In *Germany* there are lignite beds (a form of brown coal) which contain large numbers of fossil plants, animals, and insects from various regions and climates of the world. The detailed structures of these animals have been remarkably preserved and are striking proof of sudden burial. Huge diatomite deposits near *Lompoc, California* contain millions of beautifully preserved fish, usually in positions indicating sudden death. Masses of fossilized fish have been found washed together and suddenly buried in *Scotland*, while 70 tons of dinosaur fossils were found at a 13,000 foot elevation in *China*. How did they get buried so high?

> These evidences are exactly what one would expect to find after a worldwide flood like the one recorded in the Bible. They are not a result of natural death over millions of years. There is not a single example of a massive fossil graveyard

with large varieties of animals *currently forming anywhere* in the world. The rapid burial of life by the worldwide Flood of the Bible is the best explanation for the geological and fossil observations we see in the world around us.[76]

Scripture says: "Fifteen cubits upward (22-23 ft., RJ) did the waters prevail; and the mountains were covered. And all flesh died that moved upon the earth . . . every man . . . All in whose nostrils was the breath of life, of all that was in the dry land, died. . . and Noah only remained alive, and they that were with him in the ark" (Genesis 7:20-23). Thus, the real geological dynamic and panorama, from at least part of the past: *Fast and Furious*, not slow and gradual (uniformitarianism).

The evidence is strong that diamonds, crude oil, coal and fossils were formed recently and rapidly. Many well-done and informative books and research articles have been written or published on this subject, but again, you probably didn't see them in Science, Discovery or National Geographic magazines or your school, university or even public libraries because of blatant censorship. The earth seems to be *thousands* of years old, not millions or billions, including dinosaur bones. There are actually many dating methods, but surprisingly *very few* that point to possible old age for the earth and universe; most indicate that our earth is quite young. **If the earth is *young* in age then the *evolution idea is false* and creation becomes the only logical answer**.

I remember over twenty years ago when ardent evolutionist Carl Sagan, of *Cosmos* fame, who was literally infatuated with the possibility of life on Mars, was challenged in the middle of a TV appearance about the evolution idea by Kathy Lee Gifford's penetrating question, "But what about The Second Law of Thermodynamics?!" The Second Law—one of the most proved Laws of Science— also known as the law of increasing entropy, basically says that everything goes from order to disorder, from complex to simple. In other words, essentially everything changes, disintegrates, rusts, grows old, runs down towards randomness or chaos, or dies—a breakdown of complexity—no known exceptions. This is **just the *opposite*** of what the evolutionism idea calls for, which is simple to complex. Dr. Sagan, probably the leading spokesperson for evolution in North America then—if not the world— was dumbfounded by the astute question, and Regis, Ms. Gifford's popular and loquacious co-host, dropped his jaw nearly to the floor. Strong Christian Kathy Lee had done her homework;

---

[76] Richard and Tina Kleiss, *A Closer Look at the Evidence*, Search for the Truth Publications, Midland, MI, 2018, FOSSIL RECORD entry, February 13

the foundational no-exceptions-to Second Law is in direct opposition to the *Theory* of Evolution. After a stunned, silent stare, the world-renowned science pundit Dr. Sagan finally put forth his weak response: "Well, all you really need is enough time, and anything is possible." My immediate response while looking at the TV, after a chuckle and an intense stare at the *expert*, waiting for a real answer that never came: "*C'mon man*, is that the best you've got?!"

Geophysicist and seismologist Dr. Ker Thomson, former director of the U. S. Air Force Terrestrial Sciences Laboratory and Baylor University professor, logically points out the following:

> Is there available a scientifically observable process in nature that which on a long-term basis is tending to carry its products upward to higher and higher levels of complexity? Evolution absolutely requires this.

> Evolution **fails** the test. The test procedure is contained within the second law of thermodynamics. The law has turned out to be one of the surest and most fundamental principles in all of science. It is, in fact, used routinely in science to test postulated or existing concepts and machines (for instance perpetual motion machines, or a proposed chemical reaction (for viability). Any process, procedure, or machine which would violate this principle is discarded as impossible. The second law of thermodynamics states that there is a long-decay process which ultimately and surely grips everything in the universe that we know about. The process produces a *breakdown of complexity, not its increase*. This is the *exact opposite* of what evolution requires.

> The argument against evolution presented above is *so devastating in its scientific impact* that, on scientific grounds, evolution would normally be**immediately rejected** by the scientific community. Unfortunately, for the preservation of truth, evolution is not adhered to on scientific grounds at all. Rather, it is clung to, though flying in the face of reason, with an incredible, fanatical, and irrational religious fervor. It loudly

claims scientific support when, in fact, it **has none** *worthy of the name*."[77]

Back to my *"c'mon man!"* response to Carl Sagan's wimpy, unscientific, fallback reply of "with enough time, anything is possible." That's not even a legitimate statement, because relative to the foundational 2nd Law of Thermodynamics, **the greater the time span-the greater the disorganization and chaos!** Please understand that the last foundational precept—the last stronghold—that the evolutionist or atheist is *ever* going to capitulate on is the belief in long ages and billions of years, no matter what the evidence indicates. Long, unimaginable eons of time are absolutely essential for the evolutionary worldview to be even remotely feasible. The reality of a recent worldwide devastating flood would wipe out the make-believe, white lab coat-contrived *geologic column* icon in so many of our censored textbooks and museums—devastating the entire concept of evolution. (That's why the preposterous "local" or "tranquil" flood idea is promoted by *deep time* advocates.) After drilling for oil all around the world, now going several miles deep in the "fracking" industry (I just visited some sites in Ohio recently), we still have *never* discovered this mythical geologic column portrayed in all the textbooks. It **does not exist anywhere** in the world.

It is indeed a challenge to overcome the omnipresent long-ages (*deep time*) indoctrination. Just a few days ago glancing at a nature/ecosystem NOVA show on PBS, I counted the unjustified, overused "millions of years" phrase confidently uttered eight times in eight minutes! A few days later, just by chance I clicked on *Expedition Unknown* with Josh Gates on Discovery Channel. After 4-5 minutes of nothing but repeated, no evidence for, millions-of-years claims, I turned it off. These ubiquitous *deep time* claims are totally unsubstantiated. 'But, Dr. J, I hear them a lot; they must be true, right?' **Absolutely not**; there is no proof for those claims anywhere, and they're based on unprovable assumptions about the past. These radiometric dating methods have been proven to be unreliable and extremely inaccurate.

What is obvious even to a non-indoctrinated sharp fourth-grader, when presented with *all* the evidence—the stuff not permitted in censored schools and textbooks—is that these fossilized life forms in the rock layers were buried recently and rapidly, confirmed even stronger by multiple *unfossilized* dinosaur discoveries in the last two decades. What the rock record really shows is **the destruction of all life in one age** . . . not the destruction of life gradually over many, many long ages. Parents need to

---

[77] Dr. Ker C. Thompson, *In Six Days-Why Fifty Scientists Choose to Believe in Creation*, John F. Ashton, PhD Editor, Master Books, Green Forest, AR, 2000, p. 216-217.

understand this stuff and get it right, because science is probably the single most important subject in shaping a child's worldview other than the Bible.

For well over a century, and up to the present day, even many well-meaning Christian pastors and seminary professors have compromised and accepted these evolutionary/billions of years ideas and taught them in various degrees. One has to wonder why. Initially, this new worldview was promoted and catalyzed in America by liberal Unitarian Charles Eliot, who began his long tenure as President of Harvard University in 1869, already a committed evolutionist when returning from studies in Europe. This included exchanges and conversations with Darwinian zealots Charles Lyell, Thomas Huxley, and Ernst Haeckel. Haeckel, a popular German zoologist-philosopher, was one of Adolph Hitler's philosophical heroes. He falsified his still-famous embryonic drawings to promote his deceptive "ontogeny recapitulates phylogeny" evolution idea. Haeckel entertained thousands across Europe with his popular theatrical presentations promoting Darwinian evolution. Eliot's 3 main disciples—who I think of as the original Great Triumvirate of deleterious educational and cultural change in America—were botany professor Asa Gray, science and history teacher John Fiske, and Christopher Columbus Langdell, who became Dean of Harvard Law School in 1870. Asa Gray was Darwin's top promoter and public relations man in America. These Harvard professors all enthusiastically and energetically participated in the evolutionary secularization of this once outstanding Christian college.

I have on my desk a copy from the late 1800s of *Darwinism, And Other Essays*, by John Fiske, with the dedication at the beginning of the book to Englishman Thomas Henry Huxley, Darwin's unofficial—and sometimes aggressive—public relations man. Even in the 1870s—just some fifteen years after Darwin's famous *Origins* book was first published—Fiske's worshipful, religious dedication to Darwin and the altar of evolution is astounding. The rapid, almost instantaneous acceptance by Harvard's leaders of this new Christianity-crumbling radical, secular evolutionary worldview and non-scientific paradigm (with no real scientific evidence) is still stunning. In the introductory preparatory note of the second edition of Fiske's book (1885), he refers to the two most treasured friendships of his life—those with Huxley and Herbert Spencer. Spencer, bitterly anti-Christian, was profoundly important in the founding and development of misguided, evolution-based modern secular *sociology*, and the unfortunate philosophy and horrific implementation of *Social Darwinism*. Harvard history and science professor Fiske wholeheartedly supported this, which has led to incalculable human suffering and tens of millions of deaths, with Adolph Hitler's *final solution*, which included the extermination camp murders of six million Jews being a particular egregious example. Fiske's comrade Spencer, who was strongly influenced as a young man by the teachings and philosophy of Erasmus Darwin (Charles Darwin's grandfather), was another English polymath and perhaps the most well-known European intellectual in the closing decades of the 1800s. He was an ardent evolutionist even before Darwin himself, and was the expert who gave us the phrase *survival of the fittest*. Following this dedication in his 1880s book to his two best atheistic friends, in the next paragraph Harvard's influential John Fiske states this:

> We may rejoice that the time has come at last when a man may abandon old mythologies and devote himself to the disinterested pursuit of science . . . [78]

Fiske's book is full of unsupported claims, conjecture and false assumptions, with those "old mythologies" he referred to being religious belief and particularly Christianity. Nevertheless, with Harvard's prestige and status, other colleges soon followed. **Secular evolution was on a meteoric rise, and biblical history and morality began its freefall.** Other reasons for compromise by *Christian* leaders and educators did *not* include scientific evidence; there was none. It was a combination of white lab-coat intimidation, intellectual laziness, peer pressure and the residual effects of German *higher criticism*, which was an unwarranted reinterpretation of biblical history. This was spearheaded by Haeckel contemporary and evolutionist Julius Wellhausen in the late 1800's, but is now discredited (see Moses, Julius Wellhausen Ch 33, in Part III).

The issue here is not just what secular academic elites spout from their lofty perches surrounded by the smell of old vellum, or even just what the Bible says, but what are the *real facts*? Then again, that often doesn't really matter either to Harvard pundits. Their biology professor George Wald, who shared the 1967 Nobel Prize in biology/physiology, shared an interesting origins opinion decades ago (in *Scientific American*, May and Aug. 1954). Wald confessed to *only two* real options on the origin of life, one being spontaneous generation-the origin of life from non-life. He admitted Louis Pasteur (a strong Christian) scientifically disproved that idea in the 1800's (thus it was established that life *only* comes from life, the *Law of Biogenesis*). Then Wald said the *only other* possible option on the origin of life was a supernatural creative act of God. So what did the Harvard professor believe? Which option did he go with? Wald's own words: "I will *not* accept [supernatural

---

[78] John Fiske, *Darwinism, And Other Essays*, Houghton, Mifflin and Co., Boston and New York, 1896, Prefatory Note, p. vii.

creation] *philosophically,* because I do not want to believe in God. Therefore, **I choose to believe in that which I know to be scientifically impossible—** spontaneous generation arising to evolution." It's sometimes just head-shaking, although I'll give grudging respect to his honesty.

And keep in mind that even paleontologists only have the present. Those bones and materials don't have age tags on them; they're found and exist *in the present.* Most atheists, evolutionists, and secular fossil sleuths operate on the false uniformitarian assumption that the present is the key to the past. In reality, just the opposite is true: ***the past is the key to the present.*** What is really revealed by Carbon Dating, multiple recent unfossilized dinosaur discoveries, stunning human genome and genetic observations, the irrefutable R.A.T.E. (Radioisotopes and the Age of the Earth) study with helium diffusion, the lack of transitional forms in the fossil record, the absence of stellar evolution (new star formation), and the steady decline of the earth's magnetic field? What can we learn or surmise from probability and information science, irreducible complexity in living organisms, astounding ancient architecture and the over 200 different cultural stories from around the globe of a world-wide flood? These issues are directly and indirectly related to today's culture war, and how young people live their lives, and even American manhood. Let's continue our truth search.

15

NATURAL SELECTION AND *MICRO*EVOLUTION LESSON: YOUR ANCESTORS WERE NOT POND SCUM

THESE TWO TERMS, along with *adaptation* and *variability,* can be

a bit confusing to the lay person. Relative to the origins debate, they are virtually interchangeable and synonymous. We need to clarify and define the terms of the debate; sometimes they are purposely used in misleading ways. *Micro*evolution refers simply to small, minor or horizontal changes within a *kind*, or animal type. Such minor changes—variations or adaptations—are common but **never lead to a completely new body style or different type of creature**. The descendant is still clearly and always the same type—or *kind*—as the predecessor. These minor changes or variations occur by slight recombination of *existing* genetic material within the group or type. **No new genetic information is ever added**, which is an absolute requirement for imaginary onward and upward *large-scale* evolution. Size, coloration, and minor characteristics may fluctuate, but after such often-observed minimal changes cats are still cats, dogs are still dogs and finches are still finches. We see this all the time. So-called *natural selection*, which Bible-believing young-earth creation scientists accept and believe in, as well as secular evolutionary scientists, can act as sort of quality control filter which can disfavor or eliminate animals with major problems or harmful mutations. Such progeny are less fit for survival. Biblically, God designed creatures to vary and adapt to changing environments, but *not* to change into entirely different and new kinds.

Natural selection's well-known first cousin, *artificial* selection, is seen commonly throughout history and the present, such as when horse or dog breeders produce a new horse or dog breed. There are *no known* examples, though, of adaptation or natural selection, or even artificial selection (selective breeding) producing a new category of animals, or a new organ or a significantly new and different body structure—like a whale from a land mammal (what many refer to as *macro*evolution)—even over immense time periods. Due to environmental changes, stresses, and other factors these small variations or examples of microevolution—like feather coloration or beak size in birds—may occur or fluctuate, but they can often revert back to the original color, shape or size with appropriate climate or environmental changes. This has even been observed with Darwin's famous Galapagos finches—one of the most well-known *icons* of evolution.

Augustinian monk Gregor Mendel's simple but profound breeding studies on inheritance in peas in the 1800s established definite *limits* to genetic change. These well-done studies were thus ignored by the aforementioned evolution pundits and die-hards in the late 19th century. The Bible repeatedly says this same thing, about all plants and animals throughout Genesis chapter 1, stressing their uniqueness and reproductive integrity with the phrases "after his kind" and "after their kind." Beware of biological propaganda and white lab coat intimidation, and DO NOT BE MISLED: **small changes do not lead to large changes, and microevolution does not lead to macroevolution. In fact, it is absolutely impossible without the input of new genetic information—and this has *never* been observed—even though it should be frequent if evolution is true. Consequently, natural selection is, in actuality, in *opposition to* evolution.** Every geneticist and biochemist worth his salt understands that natural selection **only works with information *already present* in the genes.** So-called *natural* selection is merely the *sorting of* existing information, **then the elimination of genetic options with the concomitant death of living organisms, the very *opposite* of what evolution requires. Charles Darwin himself never cited or presented *even one* valid example of the origin of a completely new animal type by natural selection, or any other process or mechanism. Ironically, in his famous and culture-changing *Origins* book, two important subjects he surprisingly *never* talked about: 1) the actual *origin*, or beginning of life, and 2) what a *species* actually *is*—the two most important words in the very title of his famous tome!** I know it's hard to believe, and you can go ahead and waste a li'l chunk of your life and read Darwin's book if you doubt this, but it is true. You can't make this stuff up!

Despite widespread academic indoctrination and endorsement, **evolution has *never* been observed in the present world or in the fossil record or in the laboratory, and the lack of a *mechanism* for new programming addition (biochemical information) to the genome is well-known, but not talked about much by evolution advocates.** The dearth of transitional (or intermediate) fossil forms in the widespread sedimentary water-deposited rock layers around the earth has already been discussed. The many different animal types found in the fossil record **appear *suddenly*, are distinct and fully formed—with *stasis*—i.e., unchanging, with no gradualistic transitions**. Instead scientists see **huge gaps between the different types.** This strongly points to **original creation—organisms fully formed—and reproducing only "after their kind"** (a phrase repeated *ten times* in the first chapter of Genesis), in addition to their undeniable death and burial in a global mud flow/ water catastrophe. Furthermore, evolution's **last** foundational worldview pillar—**deep time**—is irrelevant also. Setting aside the legitimacy and veracity of questionable radiometric dating methods, deep time works **against the *simple-to-complex* evolution idea anyway.** As earlier mentioned, this is because the laws of science—particularly the 2nd Law of Thermodynamics (the Entropy Law) —shows that **the greater the time span, the *greater the disintegration and chaos*.** This information is not absolute proof for Creation and the Genesis historical record, but combined with other findings discussed in this book (and many exciting current research studies by creation scientists) **it is a massively huge *mountain* of evidence.**

Perhaps you can sense the obstinate mindset of evolution aficionados from the words of British atheist and anthropologist Sir Arthur Keith, an open proponent of scientific racism and one of the leading evolutionists of the 20th century, in an infrequent moment of intellectual honesty:

> **Evolution is unproven and unprovable.** We believe it because the only alternative is special creation, and that is unthinkable.[79]

Notable in Keith's well-known book, Evolution and Ethics (1947), he shares his contention that man's *moral nature* is also simply a product of evolution, an ungrounded view held and promoted by others like Thomas Huxley, John Dewey, Michael Ruse, and Harvard sociobiologist Edward O. Wilson. These famous atheists claim that the cruel and bloody evolutionary process over hundreds of millions of years not only somehow gave us mankind from apes, but an accompanying moral code. Again, you can't make this stuff up, but they have (and got paid well for it). Ruse and Wilson insinuate that Nature has *fooled us* into behaving morally (at least part of the time) for survival value, stating: "ethics as we know it is an illusion fobbed off

---

[79] Sir Arthur Keith, *Origin of Species*-100th anniversary ed., 1959, from the Forward (written some five years previously)

on us by our genes to get us to cooperate."[80] Hopefully you astute young readers and truth searchers will come to the realization that there is no evidence whatsoever to support this outrageous claim that morality is a product of evolution.

Perhaps they snatched this from the same gag-gift barrel where they originally found 1) the false idea of supposed *transitional* fossil forms in the sedimentary rock layers, 2) the totally unsupported (*never* observed) idea of new, *information-adding* positive mutations to the genome, and 3) the claim that with *deep time*, anything is possible (**not true**: the greater the time, the greater the chaos). It is head-shakingly unbelievable. It would be comical if not for so many young lives negatively affected, and the tragic results through applied, or *Social,* Darwinism.

The above moral nature/evolution-trickery view is in opposition to the biblical view. Not only did God make mankind in His own image, after His likeness, male and female, with dominion over the rest of his creation (Genesis 1:26- 28), but this included an eternal spirit capable of personal fellowship with Him.

I know many of you have thought, or asked, as I once did: "But evolution and millions of years seems to be in almost all of the science textbooks, natural history museums, National Geographic films and magazines, a lot of movies and TV science shows, and even on some national park signs and posters . . . it's got to be true . . . doesn't it?" Censorship and misinformation abounds—be smarter than that—and be smarter *sooner* than I was. Don't be intellectually lazy; be a truth detector . . . after all, this is your life . . . with eternal ramifications.

---

[80] Michael Ruse and Edward O. Wilson, "Evolution and Ethics," *New Scientist* 208 (Oct. 17, 1985): p. 50

16

GENETICS, ARCHBISHOP JAMES USSHER, THE RAINBOW, AND CAIN'S WIFE

CONCERNING BIOLOGY, BIOCHEMISTRY, genetics, embry-

ology, information science and so many other disciplines we now know so much more than in Charles Darwin's time, including the amazing complexity of a human cell, which Darwin and many of his colleagues imagined as merely a tiny amorphous gelatinous mass. *Genetics* didn't even exist as a scientific term in 1859, when Darwin's famous *Origins* book came out, and genetics as a separate and important discipline of scientific study was many decades away in the future. Unknown then were all the incredible proteins, enzymes, co-factors, trace mineral activators, and the nature and complexities of the endocrine (hormonal) system—where the hormones play a key role in the action of genes and enzymes and much more. Also yet to be revealed were the tiny *nano-machines* and cellular organelles that reek of intelligent design, along with the phenomenal cell membrane and how they all work together, akin to a perfectly designed mini- manufacturing plant.

Even more important and impressive, in every one of our 30-50 trillion or more cells, is the awe-inspiring design and coded information of the human genome-the DNA (deoxyribose nucleic acid)- and how it largely controls and directs life processes. There is no conceivable way this complex, stunning code-controlled chemical and biological manufacturing system could have developed by chance or natural processes. Any pundit or scientist who claims otherwise is dishonest, misinformed, or absolutely dominated by blind commitment to incorrect and unprovable assumptions about the past. Even chronic evolution proselytizer and atheist Richard Dawkins, from the beginning of his book *The Blind Watchmaker*, admitted the following: "Biology is the study of complex matters that **appear to**

have been designed purposely." A particular sharp young student of mine would say . . . *duh-hh*!

It has now been discovered that the sequence of coded information (nucleotides) on the DNA molecule is not necessarily just linear, or only utilized in a repetitious linear direction, but is versatile and poly-functional, and can be *read* forward *and* backward, overlapped, or turned off and on depending on need and conditions; in other words, the cell can modify its own gene expression (related to the fascinating study of epigenetics), which is absolutely, stunningly, awe-inspiring . . . and more complex than anything designed by Silicon Valley. Even after years of study, the ethereal micro-biochemical DNA software program and process in action is simply so profound, that picturing it in the mind is still challenging and evanescent at best. As science researcher and writer Brian Thomas points out, this unique DNA biomolecule also manifests the optimum characteristics for recording and preserving the *most* information in the *smallest* space, while keeping that information accessible with a double spiral structure that unwinds, separates, rejoins, and winds back at high speeds.[81] This **is by orders of magnitude the most profound information system ever discovered, far more complex than anything made by man.**

Consequently, in this 21st century all of the already mentioned scientific disciplines, and others like paleontology, anthropology, and comparative anatomy, take a back seat—*way in the back* —**to genetics** in the all-important creation/evolution debate and even the age-of-the-earth question. This is all explained and expanded on in a very understandable way—particularly for non-scientists—by cell biologist/geneticist Dr. Nathaniel Jeanson in his excellent small book, *Replacing Darwin.* As Dr. Jeanson points out, "species are defined by their heritable characteristics (traits). This means that the field of science dedicated to the study of inheritance—the field of genetics—is the most important field of science to Darwin's central question."[82]

With our current knowledge of DNA and RNA, the complexity of a human cell, genetics and the human genome, including the deleterious human mutation rate from generation to generation, the odd idea of Darwinian evolution would be **laughed out of the science laboratory today if introduced for the first time.** Dr. Jeanson refers to the sluggish development of the genetics field, but with the eventual discovery of the

---

[81] Brian Thomas, M.S., *Optimization in Creation,* Acts & Facts Magazine, Institute for Creation Research, June, 2013, p. 18

[82] Nathaniel T. Jeanson, PhD, *Replacing Darwin, Made Simple*, Answers in Genesis, USA, 2019, p. 7

unique DNA molecule of heredity in 1953. As he touches on the important collection of genome samples of many species from around the globe in recent years, he succinctly sums up Darwin's dilemma: "Darwin tried to answer a fundamentally *genetic question* long before its time."[83]

Where did all the biological coded information come from? Over 20 years ago brilliant American-Israeli physicist and information scientist Dr. Lee Spetner, in his book *Not By Chance-Shattering the Modern Theory of Evolution*, clearly showed that information *never* comes from randomness, and that there is no evidence anywhere that random variations shaped by natural selection can lead to the major changes evolution requires, regardless of how many millions or billions of years you have. It seems preposterous that **political correctness** had even strongly infiltrated Dr. Spetner's somewhat esoteric genetics/physics/information science discipline prior to 1997, when his well-documented book came out, as he issued this amusing apologetic:

I make no apologies for calling a human being a "higher animal" than the silkworm. Conventional political correctness tends to make some biologists go out of their way to avoid scaling living things into "higher' and "lower" forms. They try to avoid any value judgement by putting ourselves at the top. This avoidance often leads to awkward and unnecessary circumlocutions; I prefer to avoid the tyranny of political correctness. When silkworms start classifying living things, they are, for my part, welcome to put themselves at the top of the ladder."[84]

Over 50 years ago MIT bioengineer and mathematician Murray Eden, a specialist in *formal* languages and information theory, stressed the same thing when discussing the coded language in our DNA. He stressed that no known formal language can tolerate *random* changes (mutations) in the symbol sequence which expresses its sentences, as meaning is almost invariably destroyed. German engineer, physicist, and information theory expert Dr. Werner Gitt, in the very beginning of his powerful book, *In the Beginning was Information*, stresses how information is a fundamental

[83] Ibid, p. 7

[84] Dr. Lee Spetner, Not By Chance, The Judaica Press, Inc., New York, 1997, footnote, pp. 27-28

foundational entity on equal footing with matter and energy, specifically in *living* things, and asks: "How does information arise? How is it encoded? What is the *source of* information in living organisms? . . . the activities of all living organisms are controlled by programs comprising information . . . [it] is an essential characteristic of all life. All efforts to explain life processes in terms of physics and chemistry *only*, will always be unsuccessful. **This is the fundamental problem confronting present-day biology, which is based on evolution.**" He then stresses repeatedly that programming, or information (in living organisms), is always required for the start of any controlled process, "but the information itself is preceded by the Prime Source of all information" [God], as he refers to John 1:1-3.

It is interesting that even famous linguist and evolutionist Noam Chomsky, formerly of MIT and an admitted Marxist—not a Christian—said that **matter absolutely *cannot* produce a non-tangible entity like information (the genetic code).** And have you considered the taxpayer-funded, wasteful *SETI* program-the *search for extra-terrestrial intelligence* (or life)-which has been going on for many decades? Billions of dollars have been spent on the construction of huge radio-telescopes and the space program in an attempt to detect signals—coded information—from outer space. I can even remember being excited as a boy when visiting an impressive radio-telescope site tucked away in the hills near Green Bank, West Virginia. The point is, in this ultra-expensive and questionable program, they are looking for signals—*coded* information—knowing that such a discovery would give evidence of life, and therefore of design—or *intelligence*. Is it not stunning that many such *educated* people can look at or read about the awe-inspiring complex DNA code-language, and **not recognize** design and intelligence? It's simply mind-bogglin'.

Let's go briefly back to your genes (segments of your DNA), information, and also its transfer. For many decades now mutations have been promoted as the *mechanism* or key molecular biochemical process of evolutionary change. Mutations, though, are simply copying errors or mistakes that happen in gene/information transfer from generation to generation (when the egg and spermatozoa come together), and **involve a *loss of* genetic information**, and have *never* been shown to provide long-term benefits. **Mutations *never add* useful genetic information, something absolutely required for onward and upward evolution.** And it is not just the tremendous number of mutational changes required to transform, say, a single-celled organism into a hippo or a human which is unimaginable, but *the very order* in which such changes have to sequentially unfold. Years ago when I was finishing up a Bachelor's degree in biology/zoology, and just starting to question evolution, famous French zoologist Pierre-Paul Grasse' said this:

> The opportune appearance of mutations permitting the animals and plants to meet their needs seems hard to believe. Yet the Darwinian theory is even more demanding: A single plant, a single animal would require *thousands and thousands* of lucky, appropriate events. Thus, **miracles would become the rule**: events with an infinitesimal probability could not fail to occur.[85]

This huge problem is simply ignored by most evolutionists. For example, it does not benefit an organism to develop a wing from a foreleg until there are new accompanying ligament/tendon/nervous system changes, not to mention a lighter and extremely different skeletal system. How about the necessary and significant modifications to the lungs and breathing system (for flight)? Julie Von Vett points out[86] the following:

> Even if only 1,000 beneficial mutations are needed to change one creature into a "slightly" more advanced creature, the mutations would still have to occur **in the right order**. The odds of this happening are like flipping a coin and getting heads 1,000 times in a row-i.e., 2 to the 1,000th power. This would happen once in 10 to the 301st power # of attempts. Even if every subatomic particle in the entire known universe mutated at the fastest possible rate, and had done so for 15 billion years, there would not be enough mutations or time to explain the transformation of one form of life into another. And this is a mere 1,000 changes! The transformation of amoeba to man would require millions of changes-all in the right order.

Exciting research spearheaded by Dr. Randy Guliuzza of the Institute for Creation Research has brought forth mounting evidence that is in direct

---

[85] Pierre-Paul Grasse', *Evolution of Living Organisms*, Academic Press, New York, 1977, p. 103. Mr. Grasse' (1895-1985) was a well-known French zoologist who served as Chair of evolutionary biology at Sorbonne University for thirty years, and also was an ex-president of the French Academy of Sciences.

[86] Julie Von Vett & Bruce Malone, *Inspired Evidence-Only One Reality*, Search for the Truth Publications, Midland, MI, 2011, Feb. 18th Microbiology reading

opposition to the evolutionary paradigm based on *random* mutational change. His *continuous environmental tracking* (CET) model has highlighted mounting evidence that organisms are deliberately engineered to generate targeted solutions to environmental problems "on the fly."[87] In combination with a unique computer program model of DNA by software specialist Mitchel Soltys,[88] the findings imply that organisms actively reprioritize, reshuffle, and recombine the information necessary to generate adaptive traits.[89] Put another way, the evidence amazingly indicates that organisms purposely track environmental changes to deploy appropriate adaptive responses using the same engineering principles that underlie human-engineered tracking systems.[90] The implications of these findings are enormous, as "organisms appear to be using a mathematically precise problem-solving process analogous to a human-engineered computer program in order to arrive at targeted solutions to environmental problems."[91] As earlier discussed, it is vitally important to remember that information *never* comes from randomness. I'm confident you astute truth-seekers realize this mounting biochemical evidence is not only **the antithesis to randomness**—a foundational pillar of the evolutionary idea—but points to a programmer, or *designer*.

In fact, we now know that the human genetic code is *deteriorating rapidly*, knowledge due in part to Dr. John Sanford and his involvement in the Human Genome Project, as our human DNA now contains over 5,000 mistakes, some causing serious genetic diseases. Dr. Sanford, who invented the biolistic *gene gun* process and holds many other patents, contributed significantly to the revelation that the human mutation rate from generation to generation is stunningly *higher* than previously thought. With this observation of increasing genetic *load* (deterioration), or what he calls *genetic entropy* (mistakes piling up fast in our DNA, with a surprisingly high

---

[87] Guliuzza, R. J., Engineered Adaptability: Fast Adaptation Confirms Design-Based Model, *Acts & Facts,* 2018, 47 (9): pp. 18-20

[88] Soltys, Mitchel, Toward an Accurate Model of Variation in DNA, *Answers Research Journal,* 2011, 4: pp. 11-23

[89] Gaskill, Phil B., and Guliuzza, Randy J., P.E., M.D., Blockchain-Like Process May Produce Adaptive Traits, *Acts & Facts,* 2019, 48 (5): pp. 17-19

[90] Clary, Tim, Ph.D., ICR Scientists Present Latest Research in Wisconsin, *Acts & Facts,* 2019, 48 (9): p. 16 (presented at annual meeting of the Creation Research Society)

[91] Ibid, Gaskill, p. 19

rate of mutations), he realized that extrapolating backward in time, and applying this heretofore unexpected high mutation rate—as essentially a new genetics-based dating method—strongly pointed to a maximum age for the human race of less than ten-thousand years.

This is very much in agreement with the genealogical timeline of the Bible, and its age of six to seven thousand years for the earth (and humans). This detailed biblical timeline was first researched and meticulously outlined by historian and brilliant Christian scholar, Irish Archbishop James Ussher over 400 years ago in his classic, *Annals of the World*, where he carefully came up with a Creation date of 4004 B. C. (a little over 6,000 years ago).

This newly discovered, much-higher-than-expected human mutation rate (entropy problem) and its ramifications are all meticulously laid out in Dr. Sanford's book, *Genetic Entropy*.[92] He describes why and how he came to realize what he calls the "Primary Axiom" of biology and botany was untenable and indefensible. An axiom is a concept that is not testable but is accepted by faith because it seems obviously true to all reasonable parties.[93] Immersed in high-level genetic research for decades in a scientific and secular academic culture, with the omnipresent long-ages evolutionary paradigm always hovering, he was not a Christian at the time, but was honestly looking at the new scientific and genetic evidence after years of research. That foundational *Primary Axiom* that Dr. Sanford is talking about, that has dominated scientific thought and teaching for over a century—and especially biological evolution—goes as follows: *Life is life because random mutations at the molecular level are filtered through a reproductive sieve acting on the level of the whole organism.*[94]Dr. Sanford and many other like-minded colleagues admit this is the most untouchable pillar, or *sacred cow* within modern academia. My way of paraphrasing it would be *long-ages evolution through natural selection by random mutational change*.

The possibility and evidence for a designer or God has been banned from the education system, with censorship everywhere. Naturalism, materialism, humanism, atheism (virtually synonymous terms), or especially the presumptuous first cousin *evolution through natural selection*, has been the dominant historical and philosophical paradigm in the scientific world for well over a century now. This includes its assumption of

---

[92] *Genetic Entropy*, Dr. John Sanford, FMS Publications, 2014

[93] Ibid, p.5.

[94] Ibid, p. 5

long ages (billions of years), and no involvement of a non-material force or reality (a God or designer). But could it be false? In the previous section evidence and scientific laws *against* the evolution idea were discussed. As they used to say on a TV show, "Where's the beef?" *Where is the evidence? Evolution is not* observed in nature in the present, or in the fossil record, and *is not* supported by the scientific evidence. As geophysicist/electrical engineer Dr. John Baumgardner points out, it is based on "observationally unjustified assumptions."[95] I actually favor Dr. Sanford's courageous summary: "**the Emperor has no clothes**."[96] And then, there's my popular version: "they don't have diddly-squat."

This is powerful information. If things actually devolve rather than evolve, if things degenerate (including the human genome) rather than become more complex and refined, if information and complexity never come from randomness, **then the foundation of our entire modern education system Darwinian evolution-is in question**. This degeneration, this reality of harmful mutations and the concomitant loss of genetic information, this relentless ageing, this observed reality of complex to simple (*not* simple to complex, or amoeba to man), of order to disorder— this ubiquitous manifestation of the Second Law of Thermodynamics, one of the most proved laws of science—is seen all around us, and is irrefutable.

The origins debate is *not just a side issue*. It influences nearly everything. Even among the pundits involved in the debate, there are really *only two* choices: Either the world made itself, or this amazing universe has a Creator. Remember even evolutionist Richard Dawkins openly admits the undeniable *appearance of design* in the universe. Well-known evolutionary biologist Douglas Futuyma put it this way:

> Creation and evolution between them exhaust the possible explanations for the origin of living things. Organisms either appeared on the earth fully developed of they did not. If they did not, they must have developed from pre-existing species by some process of modification. If they did appear in a

---

[95] Ibid, Sanford, p. iii

[96] "*the Emperor has no clothes*" is a phrase that originated from a Hans Christian Anderson fairy tale where there is an emperor who paid a lot of money for some new magic clothes which can only be seen by wise people, but the clothes do not really exist, and the emperor just can't bring himself to admit he can't see them, because he doesn't want to seem stupid.

> fully developed state, they must have been created by some omnipotent intelligence.[97]

I am in absolute agreement with that statement. Recognize that these two choices are radically different—many would say diametrically opposed— despite the ironic reality that opposing pundits and proponents of the two camps (or worldviews) have the *same evidence*, on possibly the most foundational and important question any intelligent person may consider in their lifetime. The difference is in their foundational starting assumptions (see, *The Problem with Assumptions,* in Part III). Incidentally, the *'ology'* part of the words geology, paleontology, and biology comes from the Greek word *logos*; the root meaning: to gather together, to enumerate, **to give a total accounting of**. It should be apparent just from the limited (and still censored) information presented on these subjects in this book that this has **not** been done in these disciplines, particularly if it refutes or opposes the materialistic, long ages-evolutionary paradigm. Students have **not** been given *a total accounting of.*

Science writers Julie von Vett and Bruce Malone additionally point out that this all logically leads to many important creation-related issues, each of which realistically has *only two* possible alternatives[98]: "Either time/ matter/energy made themselves or they were created . . . either chemicals can come alive all by themselves . . . or they were organized by an intelligent force . . . either the DNA code could write itself or this code has a code-writer . . . either people were created fully formed and fully functional as the first man and woman . . . or some non-human creature slowly transformed into the first fully functional human being . . . either a worldwide flood rapidly formed the vast majority of the fossil-containing sedimentary rock layers of the earth, or there have been millions of years of death, disease, bloodshed, and extinction upon the earth as the rock layers slowly formed." They then rightly point out that neither option can be definitively proved in a laboratory. But the important question looms: which model— which option—is *best supported by the evidence*? In actuality, it's not even close, but you need to objectively consider this for yourself. The ramifications for your faith and how you live your life are beyond profound. And for you young men, the right choice is critical for real manhood.

---

[97] Futuyma, D. J., *Science on Trial*, New York: Pantheon Books, 1983, p. 197.

[98] *Have You Considered? Evidence Beyond a Reasonable Doubt*, Julie Von Vett and Bruce Malone, Search For the Truth Ministries, 2017, (in Foreword)

In closing this chapter, it should be mentioned that from a biblical perspective, close interrelation marriage, which is a bit uncomfortable for us moderns to even think about, was not banned until the time of Moses. Even Abraham (Abram at the time) married his half-sister Sarah (Sarai at the time; God changed their names). In Moses' time God instituted restrictions for the first time on close Relatives marrying (over 2,600 years after Creation) and with genetic entropy (cumulative mutational mistakes) building up on the human genome. Because of this *increasing genetic load*, close relatives— who would likely have similar genetic mistakes—could no longer marry (because of the greatly increased chance of birth defects). Marriage of such closely related relatives would have been fine and universally accepted in the nearly 1,760 year antediluvian period, and for a millennium *after* the Flood, until banned by God because of the aforementioned genetic entropy—or accumulating genetic mistakes (see Leviticus 18-20).

Although eventually banned by God in Moses' time, the marriage of close relatives unfortunately continued on in some quarters (even among royalty) right up into the Middle Ages, and occasionally even to this day (especially in parts of the Middle East). Biblically, Adam and Eve originally had no accumulated genetic (or DNA) mistakes; they were created with perfect genes, but all things changed with their rebellion and sin and God's subsequent Curse on them and the world. But their son Cain, who was in the first generation of human offspring ever born, would have inherited probably no imperfect genes from Adam and Eve. The effects of God's Curse (penalty) because of sin would have been minimal or non-existent in those earliest days. The Bible indicates in Genesis 5:4 that Adam lived some 930 years and fathered "sons and daughters." Highly respected Jewish historian Josephus (nearly 2,000 years ago), in his historical writings of the ancient Jewish people, said: "The number of Adam's children, as says the old tradition, was thirty-three sons and twenty-three daughters."[99] The long ages of the biblical patriarchs are of course related to not only the lack of genetic entropy (harmful mutations) in their genes, but a different, wonderful and still near-perfect world (before the Flood), probably lush with gentle topography with a worldwide sub-tropical climate. There was probably a protective water-vapor canopy, with no rain as we know it today but a *subterranean* hydrological cycle (see Genesis 1:7, Genesis 2:5 and Genesis 9:13). As a side-bar comment, computer models indicate that if such a water-vapor canopy existed with its protective, climate-controlling *greenhouse effect*, the partial pressure (or percentage)

---

[99] William Whiston, translator, *The Complete Works of Josephus*, Kregel Publications, Grand Rapids, MI, 1981, p. 27

of the critical oxygen in the air would likely have been significantly higher than today's 21% (nitrogen makes up about 78% of our air), promoting optimal health and longer lifespans. These conditions likely even somewhat favored larger animal, plant, and to a lesser degree human bodies (modern environmentally-controlled laboratory experiments confirm this with plants).

If you study the genealogical line of the early biblical patriarchs from Adam down to Noah and beyond, and their long lifespans, they rather quickly dropped down to the modern lifespan level (70-90 years) in the few centuries after the Flood. This was most likely due to a drastically changed world with more harmful radiation reaching the earth's surface (with the collapse of the vapor canopy, or the waters "above the firmament"- Gen. 1:7), a different hydrological system (now atmospheric instead of subterranean), and a likely lower partial pressure (less) of life-giving oxygen. The apparent pre-Flood absence of the rainbow (Gen. 9:13) is scientific, logical, necessary and expected by such a high, *water-vapor* canopy, which would have been very different from our clouds and atmosphere today, and reaching much farther out, preventing rainbow formation. With our present thinner atmosphere, in-close clouds and our now very small *water droplets* (since the cataclysmic Flood and vapor layer collapse of Noah's day) with conventional rain, rainbows are now possible. If you doubt this description, it is understandable, but please know that we have found all kinds of *tropical* or *subtropical* plant and animal remnants and fossils—including reptiles or dinos—under the polar and Antarctic ice, and in the now harsh climates of Siberia, northern Alaska/Canada, and Greenland.

The world was undeniably extremely different in the not-so-long-ago past, and there is much tectonic and geological evidence supporting the idea that the land mass in the antediluvian world was all together, essentially in one supercontinent. Interestingly, most long-ages evolutionists *and* young-earth biblical creation scientists accept this one giant land-mass idea; the disagreement is in *the time scale*, and on the worldwide flood issue. I remember years ago, when hearing of the discovery of 40-50 ft. alligator-like fossils in hostile ice-covered Greenland, I was initially shocked. For an excellent and logical scientific discussion on this fascinating subject, consider reading "Catastrophic Plate Tectonics: A Global Flood Model of Earth History"[100] (but perhaps not appropriate for guest bathroom reading).

---

[100] S. A. Austin, J. R. Baumgardner, A. A. Snelling, L. Vardiman, and K. P. Wise, 1994, Catastrophic Plate Tectonics: A Global Flood Model of Earth History, in R. E. Walsh, editor, *Proceedings of the Third International Conference on Creationism*, Pittsburg, PA: Creation Science Fellowship, Inc.

So why the puzzled look with furrowed brow, and question about where ol' Cain got his wife? (See Genesis 4:17). Close relative, male-female marriage was okay, then; he simply married a close female relation—probably a sister or niece. Unlike now, there would have been no genetic potential yet to produce birth defects or deformities. Just thought you'd like to know. No worries; you just have to put your biblical glasses on.

17

MORE, ON PRODIGIOUS PACKING

RECENT ADVANCES IN microscopes, imaging, biochemical laboratory techniques, and the mapping of the human genome have revealed much about chromosomes, mutations, and the heretofore hidden nano-machines in our human cells, each one of which is a veritable tiny bustling city unto itself. But in the nucleus of the cell is found the unique coded information-carrying DNA molecule in the chromosomes that deserve our greatest wonder and admiration. Australian biochemist John Marcus described it this way:

> If the DNA of one cell were unraveled and held in a straight line, it would literally be almost one meter long and yet be so thin that it would be invisible to all but the most powerful microscopes. Consider that this string of DNA must be packaged into a space that is much smaller than the head of a pin and that this tiny string of human DNA contains enough information to fill almost 1,000 books, each containing 1,000 pages of text. Human engineers would have a most difficult time trying to fit one such book into that amount of space; one thousand books in that amount of space boggles the mind. For compactness and information-carrying ability, no human invention has ever come close to matching the design of this remarkable molecule.

> Amazing as the DNA molecule may be, there is much, much more to life than DNA alone; life is possible only if the DNA blueprint can be read and put into action by the complex machinery of the living cells. But the complex machinery of the living cell requires DNA if it is going to exist in the first place, since DNA is the source of the code of instructions *to put together the machinery*. Without the cellular machinery,

we would have no DNA since it is responsible for synthesizing DNA; without DNA we would have no cellular machinery. Since DNA and the machinery of the cell are co-dependent, the complete system must be present *from the beginning* or it will be meaningless bits and pieces.[101]

From the beginning . . . absolutely; anything else is unimaginable. This complete system Dr. Marcus refers to includes the DNA, the RNA, proteins and enzymes, and much more. All of these components absolutely had to be present *together, from the get-go,* for anything to work. The possibility of these precise and ultra-sensitive biochemical molecules and components gradually evolving step-by-slow step is incomprehensible. In Dr. Marcus' own words: "It is extremely difficult to understand how *anyone* could believe that this astonishing complicated DNA-blueprint translation system happened to come about *by chance.*"[102] Indeed it is, and remember: *life only comes from life.*

---

[101] Ibid, Marcus, pp. 174-175.

[102] Ibid, Marcus, p. 177

18

THE *ONE* ICE AGE:
GOT TO HAVE COLD *AND* HOT

IF PEOPLE DON'T believe the foundational *earthly* and historical things from Genesis, including creation, the worldwide Flood, and the confusion of tongues at Babel, not only are they *not* going to accept the heavenly (or spiritual) things, as Jesus alluded to in his one-on-one chat with the scholarly Nicodemus (John 3: 12), they are going to subsequently misinterpret science and history, with all kinds of negative ramifications to follow. With the plethora of extra-biblical supporting evidences for the Bible in the last century, along with the discrediting of evolutionary dogma and German higher criticism, the Bible's *accuracy* can no longer be legitimately questioned. Its *authority* is questioned. This loss of biblical authority in Western Civilization the last 150-180 years, catalyzed by uniformitarian evolutionary belief, is the underlying cause of our current culture war, and how we got in the mess we're in. Let's discuss another one of those important, but misunderstood *earthly* things, the Ice Age.

There is actually a lot of evidence for an Ice Age, including the remaining leftover glaciers and ice sheets that exist today. The textbooks, museums, and secular scientists claim there were *quite a few* ice ages, over immense periods of time. As previously pointed out, the last foundational pillar of the materialistic evolutionary worldview, to be defended at all costs, is the belief in long-*really long*-periods of time. Without many, many multiple hundreds of millions of years the evolutionary idea is not even remotely feasible. Taking a break from exercise recently, I stopped on PBS (my *evolution indoctrination channel*) while channel surfing just to see if they might be up to their usual evolution/deep time propaganda. Immediately-on a NOVA show-I kid you not-I heard the phrase "millions of years" eight times in the first eight minutes and then turned it off. I hope you readers understand that such claims are ungrounded and unsupported by any real scientific evidence, but the smooth narrator-often Richard Attenborough-repeats the phrase incessantly as if it is irrefutable fact. The problems with unreliable radiometric dating methods the evolutionists lean on, and the unprovable assumptions they're based on, are lightly touched on in this book. Based on the comprehensive and objective interpretation of the scientific evidence, including proper analysis of deep ice cores, as well as the Bible's history (especially in Genesis and the book of Job), there was likely *only one* **Ice Age**, and it was relatively recent.

It is all related to the cataclysmic Flood of Noah's day, which was one of the most important events in world history, but is denied by atheists and secular scientists and evolutionists. This is despite extensive evidence, including the uncountable billions of fossils found in the sedimentary rock layers around the world, undeniably buried alive in a powerful water and mud-flow event. A key truth: if you deny the worldwide flood, which would have occurred some 4,300 years ago, you misinterpret almost everything that follows. The importance of *not* disconnecting your spiritual (or heavenly) beliefs from the physical world around you needs to be repeatedly emphasized.

In Genesis 7: 11-12, relative to the beginning of the Flood, it says that not only "the rain was upon the earth forty days and forty nights," but the *same* day that downpour began, "*all* the fountains of the great deep [were] broken up." Tremendous, virtually unimaginable volcanic and tectonic forces would have been released, not only with possible hyper-canes, super-tsunamis, and greatly warmed ocean waters, but also the tremendous release of sulfur and volcanic ash, which would have partially blocked the sun's heat. We actually have a fairly recent example of this on a much smaller scale, but still stunningly devastating, with the 1883 Krakatau (Krakatoa) eruption in the Sunda Strait between the Indonesian islands of Java and Sumatra. This explosion affected the entire world. It is estimated to have been equivalent to at least 200 megatons of TNT, about 13,000 times the nuclear yield of the 1945 atomic bomb dropped on Hiroshima, Japan. The convulsive explosion was heard well over 2,000 miles away on the island of Rodrigues near Mauritius, and in Alice Springs, Australia. Officially over 36,000 people died, but estimates run as high as 120,000. The volcanic ash and some 20 million tons of sulfur ejected into the atmosphere caused a worldwide reduction in temperature, by an average of 2.2 degrees Fahrenheit (as it blocked the sun), for five years. Europe did not even experience summer the next year. This was just some 137 years ago. Three months after the powerful eruption, the fine particulate aerosol and ash had spread around the world to the higher latitudes, causing an intense, red other-worldly evening afterglow. People nearly panicked in the northeastern U. S. as fire engines (mule-drawn) were dispatched in New York, Poughkeepsie and New Haven to do battle with the non-existent fires. These surreal sunsets continued for *over three years*.

The point: this was just a tiny localized hiccup, compared to what happened during the Flood of Noah's day. Imagine all . . . *all,* of the fountains of the great deep broken up, on the *same day.* If this is true, wouldn't we expect to find remnant geological evidence of that catastrophic event? Of course we would—and we do. Start by checking out a map of the earth's tectonic plates and fault lines, and the mid-Atlantic ridge on the sea bottom running all the way down the center of the Atlantic Ocean floor. Learn about turbidites and folded mountains. Yes, folded mountains. Throughout the high country and mountainous regions around the globe, there are many examples of stacked water-deposited sediment layers which have been severely distorted and folded without fracturing. The only logical explanation: these mountains were uplifted while the sediment layers were still pliable (soft) and mud-like, and with tectonic plate movement. No other explanation makes any sense at all. Evolutionists don't like to talk about this. Such geologic conditions would be expected as the earth started to settle down a bit from Noah's Flood, but with some continuing tectonic plate and land mass movement still occurring, as the earth uncomfortably dealt with its own version of acid-reflux and flatus. The term *continental drift* is sometimes used to describe this land mass movement. That term is inappropriate in light of the overwhelming scientific and historical evidence for a recent global catastrophic (*fast and furious*) mud-flow and geologic event; it was more a *continental sprint*. An example would be the Indian subcontinent slamming into the larger Asian landmass, squeezing and wrinkling-up the sediments to form the now imposing Himalayan Mountains.

Like Jesus said to Nicodemus (John 3:12), if you don't believe the *earthly* things (which scientific and historical research, and ancient records can essentially prove), you're not going to believe the heavenly (spiritual) things, like salvation and eternal life, etc. The Bible spends almost four chapters on the Flood in Genesis alone (chapters 6-9)—even more than on Creation—and refers to it repeatedly throughout the Old *and* New Testament. Back to the Ice Age topic, here is a quote from the mind of an atheistic pseudo-scholar, totally unsupported by real science that is almost phantasmagoric. I can't even count the number of unsubstantiated claims and falsehoods:

> The truly "old-time religion" developed at the end of the last Ice Age, when the *tribe* was the largest human grouping maintaining any degree of coherence. The religion of the Old Testament is a cultural fossil held over from the Pleistocene epoch, and it reflects an atmosphere of intense intergroup competition. Petrified like the bones in a paleontologist's

cabinet, the greatest ideas of the Ice Age still can be found on display between Genesis and Malachi.[103]

Of course the so-called Pleistocene Epoch is an artificially contrived make-believe time period by evolutionists; *Pleistocene* was first mentioned in 1839 by the modern champion of uniformitarianism and unfathomably long ages—Scotsman Charles Lyell. Additionally, any "great ideas" or references to the Ice Age would have to be found between Genesis 9 (end of the Flood and beginning of the Ice Age), and about Genesis 49, when the Ice Age would have essentially been over, and in some fascinating scriptures of Job, the oldest book of the Bible (with the exception of the first eleven chapters of Genesis).[104]

The resulting heated ocean waters from all the worldwide volcanic activity, red-hot magma, and lava would have created incredible amounts of evaporation, which subsequently would have fallen as almost continuous rain and as snow (in the temperate and extreme latitudes) over the colder newly formed and positioned continental land masses. As mentioned, most secular and evolutionary geologists are in general agreement with Christian young-earth geologists that the continents were once all together in one large land mass. Have you ever carefully looked at a world map, and noticed how some of the continents look like matched puzzle pieces that would easily fit together, e.g., South America and Africa? Matching continental chemical and mineral *fingerprints* somewhat support this idea. The constant heavy snows because of the significantly warmed ocean waters could not melt in the steady cold conditions. As this regular snowfall accumulated, glaciers would have been quickly formed. The transfer of tremendous volumes of water, through ocean evaporation, air movement and then snowfall to glaciers and ice would have eventually dropped ocean levels significantly. It is estimated by scientists that this ocean level drop was as much as 300 feet or more, thus exposing land bridges subsequently utilized by migrating humans and animals in the centuries after the Flood. Two examples: the shallow water between the

---

[103] Frank R. Zindler, "Religion, Hypnosis, and Music: An Evolutionary Perspective," *American Atheist*, Oct. 1984, vol. 26, p.22

[104] For some interesting scripture verses in the ancient book of Job likely relative to the worldwide Flood of Noah's day, and the subsequent Ice Age, as well as the dispersion after the confusion of tongues at Babel (with some of the most unfortunate forced to live in caves), see Job 12:15, 20, 24-25; 14:11; 30:6; and 38:8-11, 22, and 29-30. What is actually revealed from real evidence combined with Scripture: cave men, yes; *ape men*, **no**.

northeastern tip of Russia and western Alaska (there is even a Bering *Land Bridge* National Park), and the area between the southeastern coast of the Asian mainland and Australia. Where did you think most of those Native Americans came from? They crossed the temporarily exposed Bering Strait.

As the earth eventually settled down to a tenuous equilibrium and the ocean waters slowly cooled, snowfall would have diminished (less evaporation) and the glaciers would have finally retreated to about where they are presently. As this slowly happened, ocean water levels would rise and inundate land bridges and even some coastal cities,[105] while isolating certain areas and gene pools, with one interesting eventual side effect being unique isolated animal (and plant) types. Think duck-billed platypus, kangaroo, koala bear, Tasmanian devil and an array of venomous snakes and thorny plants you don't want to even get close to (Australia).

This one major Ice Age, lasting hundreds of years, fits our observations and data better than any other model. There are even hints of it in the Bible-the ancient book of Job, for example. The Great Flood of Noah's day is absolutely the key to understanding the Ice Age. Remember, you need cold *and HOT*. Here is a simple summary of the Ice Age scenario: greatly-heated oceans; partially-blocked sunlight from atmospheric aerosols and ash; tremendous evaporation; precipitation of rain in the middle latitudes and snow, snow, and more snow in the higher latitudes; glacier and ice sheet formation; and concomitant lowering of sea levels with land bridge exposure. It would have started some 4,200-4,300 years ago, and lasted for 500 to almost a thousand years.

Julie Von Vett points out that the *one* Ice Age had different effects around the world.[106] Unlike the ice, cold, and endless blizzards in the northern latitudes, heavy rains would have prevailed in the middle latitudes. During this Ice Age, the Sahara Desert was a well-watered place. It had an extensive network of rivers and lakes supporting subtropical flora and fauna. The Sahara Ice Age art has literally thousands of artistic figures etched into rocks; depictions of tropical and aquatic animals, gazelles, cattle, crocodiles and men fishing along these ancient lakes. Geologists were surprised to find bones of elephants, buffalo, antelope, rhinos, giraffes, and other animals in this area. Also found were many bones of *aquatic* animals such as hippos, amphibians, crocodiles, fish and clams, showing that this desert was **undoubtedly once well-watered**. Satellite ground -penetrating radar revealed an old drainage

---

[105] Bruce Malone, *Brilliant*, Ice Cities and Atlantis, Search for the Truth Publications, Midland, MI, 2014, pp. 40-41

[106] Julie Von Vett and Bruce Malone, *Have You Considered? Evidence Beyond a Reasonable Doubt*, Search for the Truth Publications, 2017, Oct. 7th History devotional

network in the Sahara; several of the channels were the size of the Nile River Valley. This rain (instead of snow) in the middle latitudes during the Ice Age created many lush environments. Examples:

*The Great Basin area of Nevada/Utah, USA, with the former Lake Bonneville, which was six times larger and 800 feet deeper than present day Great Salt Lake. The former shorelines can be seen to this day on the surrounding mountains.

*Eastern Pakistan and northwest India were once a beautiful and grassy forested land, as compared to the now dried-up Sarasvati River.

*Eastern Turkey was far more lush 3,000-4,000 years ago; Sargon of Agade and his men had to hack through jungles of wild game and exotic birds. No jungles are found within thousands of miles of this arid region today.

*In South America, Lake Titicaca in the Andes Mountains had a port city of huge megalithic blocks, Tiahuanaco. It is now a desolate wasteland without vegetation located five miles from the receded shoreline. During the Ice Age, it was rainy with lush vegetation.

Even from reading the Old Testament, it becomes obvious that the area of present-day Israel and the Middle East had a different flora and fauna several thousand years ago. If the earth-reforming global Flood of Noah's day is denied, much of history, geology and so much more are misinterpreted. The evidence for such a Flood is everywhere, **except within censored science classrooms, textbooks, and museums**. There is even a *non*-Christian website that lists 258 different cultural stories of a large flood.[107] Similar to the Creation/Evolution debate, the cataclysmic Deluge of Noah's day—between four and five thousand years ago—*is not* a side issue.

---

[107] *Flood Stories From Around the World*, Mark Isaak, <u>www.home.earthlink.net/~-misaak/floods.htm</u>, updated June 21, 2006

19

AMAZING PETRIFIED WOOD . . . AND *PUZZLING* POLYSTRATE FOSSILS? . . . NOT REALLY

A POLYSTRATE ANIMAL OR tree fossil is one that extends through many (poly) layers of rock (strata), in an upright or diagonal position, as Dr. John Morris and his co-author Frank Sherwin point out in their classic and beautiful book *The Fossil Record*. Coal mines often contain polystrate trees. Several of our state and national parks out west have polystrate fossils. Some of the 346 whales and other large animals found at an amazing burial site in Peru are polystrate. There are many other examples from around the world. These fossil trees and animals extend through numerous sedimentary layers of coal, shale, and limestone. Students are usually taught the untenable view that this took *very long* ages of time, as the multiple layers were deposited gradually one by one in a marine environment. That explanation makes no sense at all. As Morris and Sherwin logically explain:

> But the once-mature tree grew on land. Even if it had been somehow submerged, it could not long have survived underwater. The tree must have been uprooted and floated to its place of burial, maintaining an upright orientation as the strata accumulated rapidly around it. The Petrified Forest of Yellowstone National Park has long been thought to "disprove" the Bible's depiction of a young earth. However, the discovery of matching tree ring patterns in trees from multiple layers, showing that they grew at the same time, put an end to that. The knowledge that the upright trees penetrate more than one stratigraphic layer, that the root ball of one overlaps that of the other, and that the trees were rafted in by mudflows and did not grow in the place where they are now found, confirms their "polystrate" status . . . If hundreds of years or

more passed between depositional events, this could not be, since a tree in the lower layer would have died and rotted away before the next layer could encase it. . . . . Fossils that transgress more than one so-called "event horizon" cannot represent a long sequence of events, as is commonly taught. Sometimes the body of a fossilized animal straddles many "years" of deposition. One can only conclude that that the deposition was rapid and that the fossil was buried catastrophically in an area that was receiving a continual supply of sediments that built up around the recently deceased–or still alive-specimen.[108]

There are fascinating and beautiful pictures of these important polystrate fossils found in some wonderful books, but you probably won't find them in public school or secular university science libraries and classes, or natural history museums, because of censorship and bias. (See them in *The Fossil Record*). Most academics and evolutionists don't like to talk about them, as polystrate fossils testify to fast and furious, *recent and rapid* catastrophic water burial. As earlier discussed, it is the false but widely taught presupposition, or assumption, of long-ages evolution—not the *real* evidence and findings of science—that lead students to accept the unsupported idea that hundreds of millions of years were necessary to account for the ubiquitous fossil-bearing sedimentary rock layers. Such polystrate fossil findings do *not* support the billions of years, evolutionary paradigm so they are usually ignored. It is another *elephant in the living room*. Again, I just want to shout, *"C'mon man!"*

Likewise, it doesn't take millions of years, as is claimed on some of our national park signs and placards in natural history museums, to make petrified wood; it just takes the right conditions. Christian scientists have known this for years, and this was clearly confirmed by secular Japanese scientists years ago. Hisatada Akahane and others studied a little lake in central Japan within the crater of the Tateyama Volcano, which is filled with steaming acidic waters (reminiscent of areas in Yellowstone National Park). Fallen wood was trapped below a mineral-rich overflow and was surprisingly found to be very hard and heavy. The wood was totally petrified with silica from the waters. This and other similar finds around the world (sometimes in coal mines) puzzle geologists locked into the uniformitarian, long-ages paradigm. In this case, it was known that the petrified wood *was just 36 years old*. The Japanese scientists subsequently lowered fresh

---

[108] John D. Morris and Frank J. Sherwin, *The Fossil Record,* Institute for Creation Research, Dallas, TX, 2010, pp. 117-119.

pieces of wood attached to wires down into the same pond water, and after just seven years it was totally petrified, with the organic wood material replaced by the mineral silica.[109] Submerging some wood in my farm pond wouldn't do it, but the steady, hot mineral-rich inflow from the bottom of the volcanic lake can do it fast. The catastrophic global Flood of Noah's time (4,300 years ago) would have likely presented appropriate conditions for this process: hot, mineral-rich waters from worldwide tectonic activity, as "all" the fountains of the great deep opened up at the same time (Genesis 7: 11). It is thus not puzzling or surprising that petrified wood and amazing polystrate fossils are found scattered around the globe. Do not be misled or propagandized. One simply needs to objectively examine the real scientific evidence, and put some biblical glasses on. Just like the formation of diamonds, petroleum, coal, and *fascinating un-fractured folded rock layers, it doesn't take millions or billions of years, it just takes the right conditions.*

---

[109] Wood Petrified in Spring, *Creation Magazine*, June-August 2006, pp.18-19.

20

*UN*FOSSILIZED DINOSAUR TISSUE-WHY THE BIG SURPRISE?

IN THE LAST two decades there have been multiple discoveries of fossilized dinosaur bones containing *un*fossilized *soft* dinosaur tissue— Indicating recent burial or relatively young age   at various sites around the world. Such a shocking discovery is *unexplainable* if they really died out 65-70 million years ago or longer, which has been taught to a billion students worldwide the last 125 years. Belief in billions of years is not only widespread in our indoctrinated culture, but teaching such *deep time* in science and other fields of study is the enforced policy in the government (public) school system and most colleges. No other possibilities, like the much shorter biblical timeline—just thousands of years— is permitted because of bias and censorship, even though most dating methods indicate a *young* age for the earth. There is an interesting scripture in the Bible (Luke 19:40) where Jesus, during his triumphant entry into Jerusalem, makes a bold claim to some of the worried Pharisees that are with "the multitude." They want Him to calm things down and essentially command his disciples and many followers to shut up, quit rejoicing, and cease praising Him and his mighty works. Jesus basically says *no way*, as He responds: "I tell you that, if these should hold their peace, the stones would immediately cry out."

This particular verse from Luke, especially from 2005 on, has become more meaningful and encouraging to many Christians, especially in light of the increasing antagonism and restrictions on Christian public expression, praise, and prayer because-of all things-*dinosaur bones*. The recent multiple discoveries of unfossilized soft tissue within dinosaur bones, with sensitive complex proteins and organic compounds still stunningly present, have been popping up at multiple locations around the world . . . in "the stones," or sedimentary rock layers. The rock layers are indeed "crying out,"[110] and to the objective and honest observer, are screaming *young earth*.

This recent phenomenon began with the discovery of still-soft undecayed connective/elastic tissue, blood vessels, and cells inside a Tyrannosaurus Rex leg bone by Dr. Mary Schweitzer in 2005, at the famous— at least among fossil sleuths—Hell Creek formation in Montana. Every sharp paleontologist and biochemist—if they're honest—would admit that this extant dinosaur soft tissue should absolutely *not be there* if the deceased T-rex was truly at least 65-70 million years old, which is the time most of us were taught that the last of the dinosaurs died out. This incredible discovery should have been a lead front-page article in every major newspaper in the world, because there is no evidence-based, scientific, conceivable way this this soft tissue, remnant red blood cells, cartilaginous tissue or still-present complex proteins could be many millions of years old. Such organic molecules tend to denature or break down very quickly in unstable, hostile, or unprotected environments, or when exposed to the elements and oxygen (oxidation). Seeing is usually believing-check out the photographs yourself.[111]

The stunning appearance of recent burial and young age was eventually— and even expectantly— challenged by some evolutionists, and even a few long-ages-believing Christian pundits (like Hank Hanegraaff, of *radio Bible answer-man* fame), as they scrambled to come up with some sort of feasible explanation to support their billions of years worldview. But their ramblings just don't hold water, especially considering radiocarbon dating discoveries, and in light of even newly-discovered intact *DNA fragments*, which are ultra-sensitive to environmental disturbance and oxidation. Some of the soft tissue included elastic tissue which would spring back into place when carefully grasped, stretched, and released. Additional jaw-dropping surprise resulted from the characteristic red color—from complex, copper-containing hemoglobin molecules—still present in the degenerated red blood cell residue within the marrow spaces.

There has been widespread, wide-eyed amazement in numerous paleontology and biology labs in recent years. To these researchers trapped and framed in a *deep time* worldview, it is incomprehensible how such sensitive, complex organic molecules could still be present after the much

[110] Bruce Malone, *The Rocks Cry Out* (3 part, 18 lesson (12 DVDs) Creation Curriculum), Search for the Truth Publications, Midland, MI, 2016-2019

[111] M. H. Schweitzer, J. L. Wittmeyer, J. R. Horner, and J. K. Toporski, 2005, Soft-Tissue Vessels and Cellular *Preservation in Tyrannosaurus rex*, Science, 307 (5717): 1952-1955.

touted 65 million or more years. Conversely, if quickly and catastrophically buried relatively recently in a massive sediment-laden water event, which was indicated by the stacked sedimentary rock layers around the sleeping T-rex of Dr. Schweitzer's (as is the geological case with nearly all other such fossils), the surprising soft-tissue finds become more understandable. When burrowing worms, insects, and other critters are unable to penetrate such strata, and oxygen and most other gases are kept out, such tissue could conceivably survive for several *thousands* of years—but not millions. Timeline-wise, the global-restructuring biblical Flood of Noah's day would have been some 4,300 years ago. The Biblical model is much supported by science. If true, then some dinos (probably young ones) went on the Ark and thru the Flood with Noah, and have only gone extinct the last few thousand years. Thus there should be at least some—perhaps even widespread—evidence of man's interaction with and knowledge of dinosaurs (more on that in the following sections).

Since Dr. Schweitzer's stunning fossil find, and her instant near-celebrity status, many others have been found, including similar soft tissue by her in hadrosaur (a "duck-billed" dino) bones, which previously had been thought to be some 80 million years old. Fossil hunter Philip Bell found a hadrosaur in 2012 with real, actual dinosaur skin still present, very likely an impossibility if the creature was even just a million years old, much less 65 to 100 million plus. A similar but much earlier and even more impressive find was made by Charles Sternberg and his sons in 1908 in the Wyoming Badlands, when they discovered a dessicated, very well-preserved, nearly complete mummified hadrosaur. Photos are still available of this; one look and your first thought: *there's no way that poor dinosaur is millions of years old*! Many others have popped up ("cried out") from "the stones" in various locations around the world in recent years. One reason: the fossil sleuths are now sawing the large bones open to carefully examine the interior tissues. Because of their secular, uniformitarian, long-ages pre-suppositions and assumptions, it apparently never previously even occurred to them to do so; it was just assumed that inside the bones was total deterioration and the absence of any meaningful remaining tissue or cells. Their mindset (determined by their ungrounded, deep time worldview): it would simply be a waste of time to look inside the bones because, well, after all—they're many millions of years old—nothing besides fossilized bone could survive that!

There is no need to list other dinosaur soft-tissue finds here. Check 'em out yourself through the Institute for Creation Research, Answers in Genesis, Search for the Truth Ministries or other sources. The point: these fascinating reptilian dinosaur bones are not millions of years old, consequently their surrounding water-deposited rock layers *are not either*. These

massive and almost unfathomable imagined time scales are also preposterous in the face of new radiocarbon-dating discoveries—even dinosaur
fossil bones have measurable radioactive Carbon 14 levels—indicating
ages of **thousands of years only**. These recent dino soft-tissue finds, not
to mention many other dating methods, indicate a young age for the earth.
Remember that fossilization is an *extremely rare and difficult process*.
Organisms must essentially be buried alive rapidly by powerful aqueous
mudflows, or they will quickly decompose or be scarfed up by scavengers:

- Just south of Australia at Fossil Bluff in Tasmania are found over a
  thousand different types of ocean creatures including whales, but
  it also contains terrestrial animals that normally shun the ocean
  life . . . like possums.
- The Green River limestone formation in dry Wyoming contains
  dead remnants and fossils of salt water critters like mollusks,
  bivalves, crustaceans and sea bass; bayou or swamp animals like
  alligators and turtles; land animals like birds and mammals, as well
  as fresh water fish. These animals come from varied habitats-how
  did they get buried together?
- Far into the heartland in Nebraska there is a 5-acre fossil site where
  all different types of critters, including dinosaurs, are jumbled up
  together in a manner indicating quick, violent death. How is this
  unusual find best explained?
- Soft-bodied jellyfish have no hard body parts, and are extremely
  poor candidates for fossilization. Dead ones quickly break down
  by the sun, are scarfed up by certain scavengers, or torn apart
  by wind and wave action. Yet millions have been found wonderfully preserved in massive sandstone beds in southern Australia.
  Only quick inundation by a massive sand-laden watery flood layer
  (which in this case **covers 400 square miles**) explains this amazing
  fossil find. There is a rock quarry in central Wisconsin, obviously far
  from any major oceans that exist today, containing **thousands of
  fossilized jellyfish**. How could gobs of gelatinous jellyfish, that are
  mostly water with no skeleton or hard body parts, even fossilize?
  This particular Wisconsin find even showed the jellyfish buried in
  seven layers spanning twelve feet. Be smarter than a fifth grader:
  these soft gooey creatures were quickly inundated with mud and
  sediment—again and again (while keeping oxygen out)—to form
  the seven layers. The logical explanation is that they **were buried
  alive by violent, raging sediment-filled waters during the global
  cataclysmic Flood of Noah's day** (Gen. 6: 17).

— There is an area of several thousand square miles in coastal South Carolina with one of the richest phosphate beds in the country. I quote from geologist and hydrologist John Allen Watson, who spent a decade researching this subject:

> Phosphate beds of South Carolina constitute a vast graveyard of mixed remains of man with land animals, notably dinosaurs, plesiosaurs, whales, sharks, rhinos, horses, mastodons, mammoths, fish, porpoises, elephants, camels, deer, pigs, dogs, sheep and cattle identical to modern forms-with their coprolites (fossilized *dung* deposits-RJ). They . . . . occur in a stratum having a maximum vertical thickness of eighteen inches! The mixing of the remains is pell-mell, and the area of perceptible occurrence appears to be vast, about 6,000 square miles. Chemists of the several fertilizer companies mining the rich deposit identified the phosphate as true bone phosphate and, thus, the phosphate was believed to be derived from animal bone, the richest natural source of phosphate. Land and sea types (Leidy 1876, p. 80) occur side by side in this vast cemetery, as if they were driven together by some external terror.[112]

Widely-known and respected Swiss-American naturalist and state geologist Louis Agassiz, before his death in 1873, said the South Carolina phosphate bed was the *greatest graveyard* he had ever known. John Watson then mentions the cataclysmic Flood of Noah's day as the likely causative event, with the subsequent formation and quick tectonic separation of the continents from the original created single large land mass. He postulates the high stratigraphical level of the South Carolina coastal phosphate beds suggest "their development occurred late in this cataclysmic interval."[113] As fossil hunter Joe Taylor stresses in the preface of Watson's book: *If man, mammoths and dinosaurs lived at the same time it would utterly destroy the theory of evolution.* Sadly, Watson then points out how because such findings were so destructive to evolutionary dogma and the

---

[112] Man Dinosaurs and Mammals Together, p. 7, John Allen Watson, Mt. Blanco Publishing Co., Crosbyton, TX, 2001

[113] Ibid, Watson, p. 8

long-ages idea, then and now, the suppression of this data has occurred. Folks, it happens all the time.

- Large whale fossil graveyards have been found, mixed with many other species, including one about nine years ago near the Pan-American highway in Chile, where the dozens of whales are interestingly oriented in the same direction (and upside down), which is characteristic of similar objects in fluid-flow events (like floods) and their subsequent deposition. The skeletons are in excellent shape and very complete, with virtually no evidence of scavenging, except by some crabs on the surface. Among the fossil finds in addition to the giant rorquals (fin, minke, and modern blue whales) were strange aquatic sloths, walrus-like dolphins, modern-day billfishes, and much more. Predictably, evolutionary scientists hurriedly came up with an explanation of a toxic algal bloom, to substitute for the obvious: *a powerful violent water-borne mud flood event*. Their story is a preposterous explanation for many reasons, including no algal cell (algae) fragments in the surrounding sediments. Always remember that the British Museum of Natural History, the Royal Society, the Smithsonian Institution and National Geographic Magazine are absolutely committed to Charles Darwin and the evolutionary paradigm . . . the real evidence is irrelevant!
- Many fossils in the sedimentary rock layers, especially fish and dinosaurs, are found in contorted agonizing poses seemingly gasping for breath (the *opisthotonic posture*); some paleontologists even refer to this as "the death pose."[114] Especially with the dinosaurs, their heads are often thrown backwards with their mouths wide open, with the long tails curled forward. It is obvious they were choking on sediment, while straining for their last breath, dying of asphyxiation.
- Almost unbelievable examples of fossil animals, including fish and marine reptiles, in the process of giving birth or even attacking and devouring prey-*snap frozen in time*- have been found in the rock record, my favorite being a large intimidating ichthyosaur giving birth, obviously preserved by sudden, rapid and powerful sediment-flow burial; no other explanation is even feasible.

---

[114] *The Fossil Record*, John D. Morris and Frank J. Sherwin, Institute for Creation Research, Dallas, Tx., 2011, p. 88

These findings along with other evidences strongly indicate *recent creation* and a cataclysmic worldwide flood event (*not local*), and the coexistence of humans and dinosaurs at the same time in the past. Most folks are so propagandized about dinosaurs and millions of years, they cannot even begin to look objectively at all the evidence, and of course there is *not a hint* of long ages anywhere in the Bible. I am very sympathetic though; I was the same way as an indoctrinated young man. The average high school student by graduation has heard approximately 2,000 times, in one form or another, *dinosaurs and millions of years*. But all of this and more support the Genesis historical narrative account. The **ramifications are mind-boggling**. The origins debate, as previously stated, is not a 'side' issue. Hopefully you young and millennial men and women, an important part of my target audience with this *Handbook*, will find the correct position. It can powerfully enhance your life choices, your truth focus, your faith and becoming the proper spiritual leader of your family. Let's wind this up with a short follow-up about the two most impressive creatures (God's opinion, not mine) in the animal kingdom, including the "chief of the ways of God."

21

DRAGONS, BEHEMOTH, AND LEVIATHAN

THE WORD *DINOSAUR* is not found in the Bible, as it did not exist until 1841, when Richard Owen coined the term to describe some new-found fossil bones of a previously unknown reptile-like animal in England. If the Biblical creation account and historical narrative in Genesis is true, the animals now referred to as dinosaurs would have been created on day six of Creation week (Gen. 1: 24-31), along with the other land animals and man, but would have been called something else by Adam and Eve, their progeny and Old Testament authors. Many feel *dragons*, which are mentioned over twenty times in the Bible and in the ancient records of many different cultures around the world, were quite possibly dinosaurs. These different cultures include some going back well before the time of Christ, from Egypt to China, and some even *well after* the early Christian era; there is accumulating evidence that not only did they know about these magnificent animals, but that they co-existed.

Dragons, or apparent dinosaurs, are depicted in carvings, cave paintings (even in Europe and the western U. S.), amulets, decorative burial stones (especially from western South America), scrimshaw carvings, as well as a few ancient temples and churches (even one from a middle-ages Irish church). This includes a wonderful stone carving of a Stegosaurus at Angkor Wat, the Hindu temple complex in Cambodia which is supposedly the largest religious monument in the world (later changed to a Buddhist temple toward the end of the 12th century). Built early in the 12th century, Angkor Wat is the prime tourist attraction in the entire country, and is the symbol of Cambodia, even appearing on its national flag. The stunning, unmistakably Stegosaurus stone carving at Angkor Wat—made in the early 1100s— it looks *exactly* like our modern renditions of the large, backbone-plated reptile! How can this be? Other discovered dinosaur art and artifacts look just like our modern renditions of dinosaurs, which are mostly based on fossil findings and comparative anatomy. How could these human artists from old have possibly known what these now apparently extinct dinosaur/dragons looked like, **if**

they had not actually *seen* them? Even chroniclers of Alexander the Great recorded and described "great hissing reptiles" kept in cages in the far-eastern area of their conquering armies (present-day India/Pakistan). China is famous for its stories, legends, and artistic depiction of dragons.

It is important to know that the *average* dinosaur was the size of a sheep, and if the *Mabbul* (Heb.), or great Flood of Noah's day really happened, Noah would have likely taken *young* pairs on the ark (reptiles grow continuously throughout life) for breeding and post-flood population re-establishment, with the specific animals God had *sent to* the Ark (Gen. 6: 18-20). Consider also that according to Scripture, there was no enmity or fear between man and animals yet (but *after* the Flood-*yes*; see Genesis 9:2). Incidentally, there are at least seven other Hebrew words used in Scripture for "flood," but *mabbul,* which indicates a unique, all-encompassing cataclysmic flood, is used significantly *only* in referring to *the* Flood of Noah's day.

For so many humans to have seen, drawn or painted, and written about dinosaur-like animals and *dragons*, these fascinating reptiles must have survived the Noahic deluge and not died out at the end of the make-believe Cenozoic era some 65-70 million years ago. Undoubtedly, they struggled in the drastically changed, nutritionally depleted post-Flood environment, the subsequent Ice Age and drastic weather changes, and eventual hunting by man. A few specific examples:

- In the well-known Sumerian story, some 4,000 years ago, the hero named Gilgamesh encountered a large vicious dragon deep in the forest when hiking to cut down cedar trees. He killed the dragon, and cut off its head as a trophy (see the Epic of Gilgamesh).
- Burial stones found in caves in Peru called *Ica stones*, because of the nearby town of the same name, show dinosaurs and man together. They are less than 2,000 years old. As is the case with many ancient artifacts, some fakes or counterfeits of the originals are out there, but these are easily identified by the lack of patina (which is an oxidative residue that takes a long time to form) in the cut grooves around and in the stone. Nevertheless, because of these fakes, skeptics-especially long-ages evolution believers-reject and invalidate all of the stones. This is unjustified as many were documented and validated immediately upon discovery, and some of the early ones were even sent over to Spain in the sixteenth century, obviously well before the dino discovery era of the mid and late 1800's.
- The city of Nerluc, in Provence in southeastern France, was renamed in honor of the Tarasque, a dragon-like monster with

long pointed horns on its head (maybe a Triceratops, my favorite) that some say was brought to bay, or perhaps befriended, by Saint Martha in the 12ᵗʰ century before it was supposedly killed by the townsfolk. A carved early Gothic column capital in the cloister of the Church of St. Trophime in Aries, from the 14ᵗʰ century, is supposedly a depiction of the Tarasque.

- In various cliffs and caves in the western U. S. (and other countries, including some in Europe) there are paintings and drawings of what we would undoubtedly call dinosaurs, occasionally drawn with, or fighting another animal, like a mammoth (see *Buried Alive*, by Dr. Jack Cuozzo). The Ute Indians of western America left some of these and also had the legend of the *thunderbird,* a large dangerous flying beast or bird-like animal that reputedly would sometimes swoop down and grab little children. There are cave drawings of this fearsome flying beast (or reptile) that very closely resemble what paleontologists today refer to as a pterodactyl, or perhaps a pterosaur. This, and others, like at National Bridges National Monument in Utah, point to knowledge or interaction with dinosaurs centuries before European explorers. Didn't you older dudes ever wonder where Ford Motor Company got that neat name for its cool spots car, the "Thunderbird"? Now you know.

- On the floor of Carlisle Cathedral in Cumbria, northern England, is a small but amazing brass relief of an obvious sauropod-type dinosaur (e.g., Apatosaurus; formerly Brontosaurus). This decorative piece of brass artwork is probably close to 600 years old . . . how did they know? Those Middle Ages English artisans didn't have our extensive fossil knowledge, modern textbooks, photography, or fancy natural history museums. They now keep it covered up with a rug. If you want to see it, you better hustle. It wouldn't surprise me if it is eventually removed; it doesn't jibe with the long ages-evolution worldview.

- Sticking with northwestern Europe, we have the famous story of Saint George killing the Dragon, even represented in a wonderful woodcut by German Renaissance man Albrecht Durer; the dragon on the flag of Wales (and used on many other heraldic devices); and an Irish writer around 900 A. D. wrote an account of a large beast that may have been a Stegosaurus, as it had a tail with "iron" nails that pointed backwards, thick legs with strong claws, and a horse-shaped head.

There are many more such drawings, carvings, and stories from around the world indicating the once co-existence of man and dinosaurs, but let us wind down this discussion with the fascinating closing section of the Old Testament Book of Job, one of the oldest in the Bible. After finally appearing to Job after all of his loss and despair, God never discusses the much-mentioned issue of the suffering of the righteous. Appearing "out of the whirlwind," He commands Job "Gird up now thy loins like a man" (real men, don't you just love that?!), and lightly rebukes Job with over 70 rhetorical questions, while focusing on the majesty of *His creation*. But then, still focusing on His creative majesty and sovereignty, God moves into His closing crescendo with a focus on **the greatest of His land animal creations**, ***behemoth***, in Job 40: 15-19; 23, and says this:

> Behold now behemoth, which I made with thee; he eateth grass as an ox. Lo now, his strength is in his loins and his force is in the navel of his belly. He moveth his tail like a cedar; the sinews of his stones [thighs] are wrapped together. His bones are as strong pieces of brass; his bones are like bars of iron. He is *the chief of the ways of God*: he that made him can make his sword to approach unto him. . . . Behold, he drinketh up a river, and hasteth not: he trusteth that he can draw up Jordan into his mouth.

"Huge beast" is the meaning of behemoth, and don't let some compromising pastor or Bible *expert* try to pass this off as an elephant, hippo, rhino or other large animal-the description *fits none* of those, but it does fit a large dinosaur (dragon?), with a tail "like a cedar," the centrally-located structural power force "in the navel of his belly," and bones like "bars of iron." It is quite probable, logical, historical, and now with even scientific support that descendants of the dinosaurs on the ark of Noah survived not only to Job's day, but far beyond, contributing to the scattered but nearly worldwide traditions and legends of dragons. Surprisingly, even from France went forth a scientific expedition to the pygmy land of tropical rain forested west-central Africa less than forty years ago, as the stories and testimonials of dinosaur-like still extant animals seemed so legitimate.

Another well-known possible ancient reptile sighting(s) is that of the so-called Loch Ness monster (locally known as "Nessie") in Scotland; the first supposed sighting was in April, 1933. More followed, and in the 1960s several British universities conducted visual and sonar expeditions on the lake, with no conclusive findings, but in each expedition sonar analysis detected some type of large, moving underwater objects. A 1975 expedition with sonar and underwater photographic equipment in Loch Ness got a photo that showed a structure that resembled a giant flipper of an aquatic animal (plesiosaur?). Other possible sightings and photographs are unverified.

In Job 41 following the behemoth account is a very detailed and lengthy description by God of **the greatest of His *marine* reptiles or dinosaurs—*leviathan*—**"the dragon that is in the sea" (Isaiah 27:1). Did you ever wonder about all the "sea monster" and "dragon" stories when you were young? I sure did, along with the sometimes "fire-breathing" part. Job 41: 21 it states "his breath kindleth coals, and a flame goeth out of his mouth." It is possible that some dinos may have had some type of combustible gas ability, akin to today's Bombardier beetles, who can eject a hot noxious chemical spray as a defense mechanism to be reckoned with. In verse 10 it says that no man is strong or "fierce" enough to handle leviathan, but God could, as was the case with behemoth (Job 40: 19).

I realize that hard-core skeptics will usually reject, or simply ignore, many of the evidences and arguments I have just shared. As retired chemical researcher and science speaker/author Bruce Malone says: "In spite of clear evidence that people have seen dinosaurs, skeptics will simply choose to believe that these types of artifacts do not represent real dinosaurs or were not produced by ancient people. Microbe-to-man evolution *is dogma rather than science,* so no amount of evidence indicating that man and dinosaurs co-existed will convince evolution believers of this fact."[115] I agree about the dogma; evolution is just a story to explain things without God. I accepted and believed it for a while as an undergraduate science major; it was the **only origins idea that was taught or allowed**. It makes you wonder what the academic powers that be are afraid of.

---

[115] Malone, Bruce, *Censored Science-The Suppressed Evidence,* Search For the Truth Publications, Midland, MI, 2014, p. 71.

22

DRAGONS (DINOSAURS) IN SCRIPTURE, AND WHAT ABOUT THE UNICORN?

IF INDEED THESE great and now seemingly extinct reptiles we moderns call *dinosaurs* were created by God, on day six of Creation week with the other land animals and man (Genesis 1: 24-27), surely the Creator would mention or talk about them in the Bible. Mankind would have co-existed with *dinos* for a while, especially before the Deluge of Noah's day—and afterwards for a while—but Biblical authors would have obviously used different terminology in describing the great reptiles. *Behemoth*, *leviathan*, and *dragons* (Hebrew *tannin*) are mentioned multiple times in Scripture, and likely refer to these captivating beasts. As you consider these scriptures, remember that these 47 brilliant *Authorized Version* Bible scholars in the early 1600s were highly qualified, and obviously considered dragons as real, historic animals that co-existed with man:

Job 40: 15-24-Here God begins to describe the greatest of all of his land animal creations, *behemoth*, after asking Job dozens of rhetorical questions, which focus on his majestic creative ability, not the suffering of the righteous. This behemoth feeds on grass like an ox, his strength is in his loins, he moves his tail like a cedar, his bones are as strong pieces of brass and bars of iron, and he is the chief of the *ways of God*. Job was familiar with the beast. In verse 23, it says he "drinketh up a river . . . and can draw up Jordan into his mouth." Compromising pastors and liberal theologians sometimes refer to this animal as a hippo or an elephant, which **is ridiculous**, as the description doesn't fit either one of them. The description fits a large sauropod dinosaur like an Apatosaurus (formerly Brontosaurus).

Job 41: 1-34- Here God describes the greatest of his created marine reptiles, Leviathan, "the dragon that was in the sea" (Isaiah 27: 1). On careful reading it becomes apparent that it was not a crocodile, shark, iguana or some marine animal we are familiar with today. Leviathan could have been a plesiosaur, or something similar. Apparently, no man could then put a hook through his nose, or fill his skin with barbed iron, or his head with fish spears, and there *was no man that dare stir him up* (Job 41: 10). His scales were so tight and impenetrable that air could not even come between them, and they could not be sundered. The great reptile could apparently expel a mixture of smoke, flame, and combustible gases (common in dragon legends, and even in today's bombardier beetle). When he raised himself up, the mighty were afraid, and he made the deep boil like a pot.

Psalms 74: 13-14-"thou brakest the heads of the dragons in the water," and "thou brakest the heads of the leviathan in pieces" (In verses 13-15, Scripture is referring to the Flood, and the breaking open of "all the fountains of the great deep").

Psalms 104: 26- "There go the ships: there is that *leviathan,* whom thou hast made to play therein."

Isaiah 27: 1- "leviathan, the piercing serpent . . . the *dragon* that is in the sea"

Psalms 148: 7- "Praise the Lord from the earth, ye *dragons*, and all deeps"

Psalms 91: 13- "Thou shalt tread upon the lion and adder: the young lion and the *dragon* shalt thou trample under feet"

Ezekiel 29: 3-"Speak, and say, Thus saith the Lord God; Behold, I am against thee, Pharaoh king of Egypt; the great *dragon* that lieth in the midst of his rivers, which hath said, My river is mine own, and I have made it for myself."

Isaiah 34: 13-"And thorns shall come up in her palaces, nettles and brambles in the fortresses thereof; and it shall be an habitation of *dragons*, and a court for owls"

Nehemiah 2: 13-"And I went out by night by the gate of the valley, even before the *dragon* well, and to the dung port, and viewed the walls of Jerusalem,"

Job 30: 29-"I am a brother to *dragons*, and a companion to owls." (This is when he had lost everything, and was banished to the wilderness)

Psalms 44: 19-"Though thou hast sore broken us in; the place of *dragons*, and covered us with the shadow of death"

Isaiah 13: 22-"And the wild beasts of the islands shall cry in their desolate houses, and *dragons* in their pleasant palaces"

Isaiah 35: 7-"in the habitation of *dragons,* where each lay, shall be grass and reeds and rushes"

Isaiah 43: 20-"The beast of the field shall honor me, the *dragons* and the owls: because I give waters in the wilderness, and rivers in the desert, to give drink to my people, my chosen"

Jeremiah 51: 34-"Nebuchadnezzar the king of Babylon hath devoured me, he hath crushed me, he hath made me an empty vessel, he hath swallowed me up like a dragon" (*Note: the referred-to dragon must have been capable of swallowing a man whole, otherwise the verse makes no sense)

Micah 1: 8-"Therefore I will wail and howl, I will go stripped and naked: I will make a wailing like the *dragons*"

Malachi 1: 3-"And I hated Esau, and laid his mountains and his heritage waste for the *dragons* of the wilderness"

It is becoming increasingly apparent that these unique reptiles existed recently, and most of them perished rapidly in a violent mud-flow water cataclysm some 4,300 years ago. We need to get this right. Again, if you deny the recent worldwide Flood, you misinterpret everything else. Dinosaurs, and indoctrination that they *evolved* and lived many millions of years ago, is probably the number one concept that has led untold millions of young people away from Biblical truth and real history. Closing Notes:

In Nehemiah 2: 13, mention of a "dragon well" is particularly interesting, and may have been named as such by the early migrants who first traveled there after the language confusion at Babel, when dinosaurs (called *dragons* before 1841), regularly visited the spring.

In Psalm 74: 13-14, the "dragons in the waters" (same as *leviathan,* or the great marine reptiles), which could not be taken with human weapons, were broken and destroyed by the powerful forces of the Flood.

The description "fiery flying serpent" is also mentioned, including Isaiah 14: 29 and 30: 6; this could refer to a pterodactyl (or pterosaur) flying reptile, with the capability of expelling a smoke-combustible gas combination, again similar to today's Bombardier beetle. Pterosaurs are likely responsible for the legends of the "thunderbird," which are found in Indian stories and artwork from Alaska to South America, including a well-known examples in a Utah canyon and a Peruvian Ica stone.

In discussing dinosaurs, extinct animals and the Bible, it should be noted that the *unicorn* is also mentioned in the closing chapters of Job (Job 39:9-10), as well as Psalm 22:21. Is the Bible saying that those one-horned, usually light-colored horses were real? Several animal terms and names have been changed (unjustifiably, in my humble opinion) in most of the newer Bible translations, like "wild ox," for unicorn. But the word "unicorn" is the term used in the older versions, including the wonderful King James (or Authorized) Version, which was first available in 1611, after careful, meticulous work for over seven years by many of the top scholars in Christian Europe. Here is the often ignored point: At that time, the rhinoceros was referred to as *a unicorn*. Well into the 1800s the unicorn was still defined in dictionaries (which actually were somewhat *new*; Noah Webster published the first one in the U.S. in 1806) as an animal with one horn, the Monoceros, a name often used for the rhinoceros. The Indian subcontinent's one-horned rhino of today is still scientifically classified as *Rhinoceros* **unicornis**. "Wild ox" is not what the original scriptures said, and the characteristics mentioned in the Bible fit the rhino: "great strength" (Numbers 23:22, 24:8); probably dangerous to children, and not suitable or trainable for plowing (Job 39:9-10); and even then, some had two horns (Deuteronomy 33:17), or just one (Psalm 92:10).

In closing, Lamentations 4:3 interestingly says this: "Even the *sea monsters* draw out the breast, they give suck to their young ones." The Hebrew word here translated "sea monsters" is, once again, *tannin*, which is occasionally translated "whales" or "serpents," but most of the time (in the KJV) "dragons." Modern Bible versions sometimes puzzlingly translate it "wolves," "hippos," "jackals," or even "crocodiles." This confusion arises from two reasons: 1) the *tannin,* whatever they actually were, are now apparently extinct, and 2) some of the modern Bible version translators were strongly influenced by uniformitarian evolutionary dogma. Dr. Henry Morris offered this:

> In the sixteenth century and earlier, however, accounts of dragons were still so widely known and reliable that scholarly bible translators saw no problem in identifying the *tannin* as "dragons." They knew that the biblical accounts correlated realistically with the many similar records in early and medieval literature. Since the first dinosaur bones were discovered **less than two centuries ago** the biblical accounts have been found to correlate with information paleontologists have provided about dinosaurs, from reconstructing the many fossils of these once-abundant animals. However, the particular *tannin* in this verse seems to be a mammal, whereas most dragons seem to have been dinosaur-like reptiles. Possibly at least one kind of dragon/

dinosaur was similar to the platypus, which has features of both reptiles (laying eggs) and mammals (suckling its young). Perhaps *tannin,* was understood as a generic term, applied to any monster-like animal. Paleontology has also revealed a number of exotic animals called mammal-like reptiles; many of these also were large and grotesque. In any case, dragons were real animals-probably dinosaurs or mammal-like reptiles or both— which did not become extinct until relatively modern times.[116]

Finally, dragon legends from around the world often portray the beasts as fierce, dangerous, and attacking human beings and then being overcome and killed. It is significant that in the Bible Satan is even described as a powerful dragon.[117] This important passage from the first part of Revelation 12 explains the sign of the great dragon. Upon serious reflection, one could make a very solid, logical argument that in order to be used as a legitimate sign, dragons must have been *real animals*, and beasts that were widely known, respected, avoided and feared by the people of old (review Isaiah 27:1, and the other dragon scripture verses). Keep in mind that the word *dinosaur* was not coined until 1841, and is a relatively new term. Only in the last century and a half, with the rise and domination of the long ages/ evolutionary worldview have scholars and critics denied any human knowledge of, and interaction and co-existence with dragons—which history, real science, and the Bible indicate were dinosaurs. If dragons are simply make-believe or myth, which is extremely unlikely when all of the evidence is considered, then the description of Satan as a great adversarial dragon is mythological as well. Dragons were dinosaurs, well-known to brilliant King James scholars, biblical authors, and people of the past.

---

[116] Henry M. Morris, Ph.D., LL.D., Litt. D., annotation from *The Defender's Study Bible* (KJV), World Bible Publishers, Inc., 1995, p. 844

[117] God, "God-breathed" thru the Apostle John, King James Bible, Revelation 12:3-9, 1611

23

WHERE'S THE DINO FOOD?

AS WE DISCOVER more and more animal fossil remains in the sedimentary rock layers around the world, sometimes even including coprolites, animal and dinosaur footprints, and even extant *un-fossilized* dinosaur soft tissues within their bones, another puzzle arises for evolution-minded paleontologists. They wonder why so many of these stratified large-animal-bearing rock layers contain little to no evidence for shrubs, trees, flowers, or fruit-bearing plants necessary for food and life support for the large critters. Besides the remnant large animal bones and fossils, the formations otherwise seem almost *sterile* of plant or botanical life forms. How can this be? If slow and gradual, or millions of years of sediment deposition occurred, as the evolutionists have claimed for a century and a half, where is the food (plants)? How did these animals survive? Many of the stacked rock layers found in the Grand Canyon and throughout the western U. S. contain various animal fossils and trackways, sometimes even *sorted* (like the millions of nautiloids in Grand Canyon's 600 feet-thick Redwall limestone layer), but no plants or trees.

I had the good fortune some years ago to participate in a 6-day rafting, camping and geology trip down the Colorado River through the breathtaking Grand Canyon. This included my two teenage sons and a wonderful group from the *Answers in Genesis* ministry, and included several PhDs which helped with the geology-fossil sleuthing and discussions. Again, evidence of quick, violent, and recent canyon formation, with millions of animal burials, but few (often zero) tree/plant fossils. The large mammals and dinosaurs found in some of the rock formations around the earth would have required a very large amount of food. Scientists estimate that the large herbivorous dinos ate 2-3 tons of plant food per day. So, where's the dino food? It's just not there.

The best explanation, supported by the scientific evidence, is that these layers were laid down rapidly and often violently during a global flood or in its immediate aftermath.[118] Observed is the characteristic sorting which often occurs in fluid-flow events, while plants matted together, were buried, and formed the coal deposits scattered around the globe. Incidentally, the earlier mentioned nautiloid deposit found in the Grand Canyon, scattered thru the lower level of the Red Wall Limestone deposit, was discovered and described by geologist Dr. Steve Austin. These really cool, sugar-cone-shaped marine mollusks were statistically oriented in a similar westerly flow direction throughout the massive layer.[119] This only makes sense if the entire massive 200 yard-thick layer was catastrophically laid down in a powerful flood and mud-flow event **all at once, trapping the estimated *one billion* nautiloids**. Their interesting arrangement, with sorting and lay-ering, is what one would expect in such a powerful, destructive fluid-flow event (confirmed by laboratory experiments). Put your biblical glasses on, and don't be further propagandized. The Flood of Noah's day was real, the evidence is all around the world, and the majestic Grand Canyon *screams* recent and rapid catastrophic formation.

---

[118] There is geological evidence supporting the idea that as the floodwaters drained off of the newly formed and exposed continental land masses (some 4,345 years ago), naturally formed temporary earthen dam reservoirs often broke with the release of tremendous volumes of water with great destructive power. This includes Lake Bonneville, which covered most of present-day Utah and parts of Nevada and Idaho, and led to the formation of the Grand Canyon. There is even evidence that this unimaginable water power flowed partially uphill, and cut through solid rock seemingly as if it were butter. Part of the ancient shore-line of this *mabbul*-caused, remnant giant lake can still be seen today in Utah. Other explanations for the Grand Canyon formation do not jibe with the evidence, and make no sense.

[119] Steve Austin, *Regionally Extensive Mass Kill of Large Orthocone Nautiloids, Redwall Limestone (Lower Mississippian), Grand Canyon National Park, Arizona*, presented at GSA annual meeting, Denver, Co, Oct., 2002. Also, an abbreviated but wonderful version by Bruce Malone, in *Censored Science-the Suppressed Evidence, Mass Extinction in the Grand Canyon*, Search for the Truth Publications, Midland, MI, 2014, pp. 82-83

24

THE MALE/FEMALE EVOLUTION ENIGMA

ALL COMPLEX (OR higher) life forms need male and female to reproduce. The evolution idea has a **huge** problem here, which evolution aficionados try to avoid talking about. Let us reason together. There is no solid evidence anywhere which supports this idea, but hypothetically—if evolution were true— which came first, the male or the female? The odds of the two genders evolving over time—but **at the same time**—through *their own particular random mutational changes* with all of the required organ, nervous, skeletal and endocrine system changes and connections, with the complementary male-female adapter match-up features, so to speak, and *in the same immediate geographical area* (with the probability of getting together), are incalculable and easily beyond the prognosticating abilities of the sharpest Las Vegas bookmakers. It is incomprehensible. With our recent newfound genetics knowledge, it is even unimaginable.

Males were created fully functional and unique with the gift of language from the get-go (see Genesis 1: 28-30; 3: 8-19)), and created to be males. Females were created differently (Genesis 2: 21-23), and diff*erent*, but fully functional and unique from the get-go, with the gift of language, and created to be females. Both also possess on an *equal* basis an eternal spirit not only capable of personal fellowship with their Creator, but additional and special mental and spiritual abilities not found in even the higher animals. These include abstract thought; the discovery of mathematical and scientific equations and laws; emotional feelings; an appreciation of beauty; creativity; written language; music language, composition and creation; moral conscience and so much more, including the amazing uniquely human capability of praising, worshipping and loving God. So what do the hard-core evolutionist/deep time believers have to say about the male/female evolution enigma? Nothing; it's another *elephant in the living room* . . . they don't like to talk about it.

25

APE TO MAN: THE GENITALIA PROBLEM

DESPITE OVER A century of indoctrination, there is no evidence that man evolved from apes. The sordid history of the search for the missing link between ape and man is, in fact, filled with outright frauds and lies, or misinterpretations (among the most well-known: Java Man, Nebraska Man, Piltdown Man, and Peking Man). There are some similarities between man and primates, but this hints more of *common design*, rather than common descent. The differences are *profound,* though, and include thumbs on the feet of apes, radically different noses and lips, different cervical spine and head morphology, and a huge difference in tongue and larynx anatomy (allowing speech in humans) combined with our unique muscles of facial expression (not found in apes).

One of the most fascinating and significant differences is found in comparing and contrasting the male genitalia. Male apes have a *rigid bone* in the reproductive organ, while male humans have a complex chemical/emotional/hormonal *hydraulic* system. There is no conceivable reason, pathway, how, or why the rigid reproductive organ bone in an ape could drastically change into the complicated usually soft hydraulic system found in humans. It is mentally impossible, even by a doctor or human anatomy professor knowledgeable in physiology, anatomy, genetics and such, to *even envision* such a step-by-step mutational process (and *not* a pleasant line of thought anyway). Apes are apes, and humans have always been humans-created in God's image. As the late Harvard paleontologist Stephen J. Gould surprisingly once said: "We're not just evolving slowly. For all practical purposes we're *not* evolving. There's no reason to think we're going to get bigger brains or smaller toes or whatever-we are what we are."[120] Or, as Dr. Watson (not Sherlock's pal) admitted in Science Digest

---

[120] Stephen J. Gould, (former professor of Paleontology at Harvard University), in an Oct., 1983 speech; as reported in 'John Lofton's Journal,' *The Washington Times*, Feb. 8, 1984

back when I was in dental school: 'Modern apes, for instance, seem to have sprung out of nowhere. They have no yesterday, no fossil record. And the true origin of modern humans—of upright, naked, tool-making, big-brained beings—is, if we are to be honest with ourselves, an equally mysterious matter.'[121]

---

[121] Dr. Lyall Watson, 'The water people,' Science Digest, volume 90, May 1982, p. 44. Dr. Watson was a respected South African botanist, ethologist, zoologist, anthropologist, and author and as a young fellow learned about nature, wood-craft and bushveld life from the eminently qualified Bushmen and Zulus.

26

NO NEW STARS?

IF THE ASSUMPTIONS of billions of years (also referred to as *deep time*) and evolution are true, we would logically expect to observe stellar (star) evolution and death as well. We definitely have observed the *death* of stars through the centuries, more keenly in recent years with the modern improvement in telescopes. When a star dies—which is a violent event—a bright explosion is sometimes seen, and is called a supernova. About two hundred have been recorded. Mankind observes such a supernova once every 25 years, and its subsequent gas and dust remnants should remain visible for millions of years. But having only observed about 200 of these supernova remnants, that calculates out to only 6,000-7,000 years of supernovas, coinciding very closely with Archbishop James Ussher's detailed calculation of the earth's (and universe's) age based on the Creation account in Genesis and the detailed patriarchal genealogies, discussed earlier in this book. Stars are seen dying regularly; we have *never seen* one forming. All observed cosmic events are destructive, *not creative*. Nevertheless, if new stars are somehow forming (evolving), astronomers should observe new starlight where none has previously been seen (it's really *not* rocket science). This has never been observed, even in any region of the sky photographed through telescopes decades earlier. Similar to other areas of science when evolutionism believers are faced with clear evidence in opposition to their worldview, stories are made up to try to explain things without a Designer (or God). Along this same thread, it is recently reported that new stars are forming in different areas of the night sky, but what is actually observed is *not new star birth* but merely great dust clouds surrounding existing or former stars. The preposterous idea often promoted and taught that stars somehow form (or evolve) from clouds of *dispersed* gas goes against the known and proved laws of science, and has *never* been observed. For you young *Trekkies* and star-gazers, former chemical researcher Bruce Malone explains:

Stars are enormous, densely packed spheres of hydrogen gas with nuclear fusion taking place inside, resulting in the release of enormous amounts of energy. Yet hydrogen gas will not condense into a super-concentrated ball within the vacuum of space for three fundamental reasons: First, gas gets hotter as the molecules are forced closer and closer together. Long before gravity could become strong enough to take over and condense nebular hydrogen clouds, the gas pressure would be a million times too strong to allow the continued collapse of the cloud. Second, the gas within the nebular clouds is rotating. As the material is pulled closer, it would rotate faster and the increasing angular momentum would drive the material apart, not allowing it to condense into a star. Third, nebular gas fields have a magnetic field associated with them. As the material within this field collapses, the field would become increasingly concentrated and the material would repel like the poles of two magnets. Thus, there is no mechanism explaining the formation of stars, while there are numerous observations indicating that stars cannot form by *any* natural process.[122]

The bottom line: scientists have *never seen* the birth of a star, which would be expected if deep time and the evolution idea were true, but have indeed observed the deaths of hundreds of stars. Astronomers now know as well that the universe is not uniform in temperature, which is strong evidence *against* the idea of the solar system and universe being infinitely old. Just look around the night sky on a clear night, or through a cool telescope: there are **hot** stellar objects **everywhere**-they *couldn't be real old*. I hate to say it again . . . but it's *not* rocket science! Do not be misled, or propagandized. Things are young.

Additionally, our modern high-tech powerful telescopes reveal congregated masses of billions of stars, which we call galaxies, not only moving super-fast through the vast reaches of space but *clustered tightly* together. It is estimated these galaxies each contain 100 billion stars. If these galaxies began as a single point of matter, as has been taught by the big bang theory (which is rife with problems), they should be uniformly dispersed by now, *not* clustered. Is it not truly awe-inspiring that these massive, almost unimaginable clusters of stars are they themselves clustered

---

[122] Bruce Malone, *Censored Science-The Suppressed Evidence*, Search for the Truth Publications, Midland, MI, 2012, pp. 86-87, and taken partly from *Taking Back Astronomy*, by Dr. Jason Lisle, Master Books, 2006.

with countless *other* such masses of stars (galaxies), flying through the universe? In explosive events, things tend to quickly spread out. Things are *not* evenly or uniformly spread out, or distributed, through the deep void of space. The galaxies within these amazing galaxy clusters are so close together, they simply could not have been flying apart for long periods of time, and gravitational attraction cannot explain it either. No new stars, plenty of star death, and clustered galaxies *scream recent creation.*

27

HOW ABOUT TIME, DISTANT STARLIGHT, AND THOSE *STRETCHED-OUT* HEAVENS?

AS MENTIONED ALREADY, when something is strongly emphasized or repeated over and over in the Bible it is noteworthy and particularly important. I would think such examples would even catch the interest, or at least pique the curiosity, of the skeptic or atheist who might be reluctantly checking the Scriptures out. In chapter 9, verse 8 in the ancient book of Job is found the first mention of *many* biblical references to God "spreading" or "stretching" out the heavens, where it says "[He] alone spreadeth out the heavens." This is actually important relative to the issue of distant starlight and time. Here are some other examples:

> Psalm 104:2-"who *stretcheth* out the heavens like a curtain,"

> Isaiah 42:5-"Thus saith God the Lord, he that created the heavens, and *stretched* them out,"

> Isaiah 40:22-"He . . . that *stretcheth* out the heavens as a curtain,"

> Isaiah 44:24-"Thus saith the Lord, thy redeemer, and he that formed thee from the womb, I am the Lord that maketh all things; that *stretcheth* forth the heavens alone,"

Isaiah 45:12-"I have made the earth, and created man upon it: I, even my hands, have *stretched out* the heavens, and and all their host have I commanded."

Isaiah 51:13-". . . the Lord thy maker, that hath *stretched* forth the heavens," and

Jeremiah 10:12-"He hath made the earth by his power, he hath established the world by his wisdom, and hath *stretched* out the heavens by his discretion."

The Bible clearly says in Genesis and later in Exodus 20:11 (also, see Exodus 31:18) that God created everything *in six days*. Hebrew scholars have confirmed that the clear meaning of the text is unquestionably regular 24-hour days. In fact, the Hebrew word for "days" (*yamim*) is used over 700 times in the Old Testament, and can *never* be demonstrated to require any meaning except that of literal 24-hour days, and unquestionably nothing close to long ages or millions of years. Whenever the word "evening" or "morning", or the phrase "evening and morning," or a numeral is associated with *day*, it always means a 24-hour regular day. It is quite interesting, and significant, that in this well-known passage from Exodus 20 where the Ten Commandments are given, the one demanding to "Remember the Sabbath day, and to keep it holy" (Exodus 20:8), is the *only one* where God takes the time to then carefully and deliberately remind his people of its meaning and explain Himself. It is almost as if He wants to make sure there is no equivocation or misunderstanding, and that He took six days—instead of one, or ten, or a single second—to accomplish his work of creating *as a model for us*. God's week was the template, or pattern, for humanity. It is fascinating and significant that our other commonly used measurements of time (the day, the month, or the year) are all keyed to astronomical processes and movements, but **not** *the week*. Certain communist/socialist countries in the past have even experimented with eliminating the conventional 7-day week, extending it to ten or more days; it has never worked out. We humans keep time in weeks because God does.

So how does one mesh God creating everything "in 6 days" (and less than 7,000 years ago), based on the biblical genealogies, new genetics findings, and solid dating methods, while considering the apparent indicators in space of vast distances and old age? And where and how do all of those biblical mentions of God originally "spreading" or "stretching out" the heavens factor in?

It's all about physics and time dilation. Brilliant Christian astrophysicist Dr. Russell Humphreys wrote the book *Starlight and Time* which essentially solves the problem of distant starlight in a young universe. Most people don't realize that time passes at different rates depending on the location where the passage of time is occurring, or being measured. For example, a clock at sea level on the west coast runs a tiny bit slower than one on top of Mt. Whitney or Pikes Peak. Dr. Humphries has proven that millions of years could have passed out in space while only six days were passing during Creation week on the earth. The mathematics show that when the substance, or fabric, or *"curtain,"* if you will (Psalm 104:2, Isaiah 40:22) of space was being stretched out, the passage of time in some areas would have stopped. At the same time, many millions and theoretically even billions of years could have been passing only moderately further out in space. Near the center of the universe, time could have essentially stood still in comparison to areas outside of and beyond the event boundary. Thus on day four of Creation week, when (according to Genesis) the stars were made, crunching the numbers shows there would have been an unfathomable increase in the mass of the universe.[123] This is heady stuff, but Dr. Humphreys explains that literally billions of years could have then passed outside our solar system, while only a single literal day was passing on earth.[124] This is the time dilation I referred to earlier. We have a lot to research and learn, but there are answers out there. *Starlight and Time* is a relatively small, highly readable book, and is strongly recommended. A deep understanding of Einstein's theories, quantum mechanics, and cosmological esoterica is *not* necessary!

---

[123] Samec, Ronald. "Explaining nearby objects that are old in time dilation cosmologies," Journal of Creation 28 (3) 9, 2014.

[124] Humphreys, D. R. "New view of gravity explains cosmic microwave background radiation," Journal of Creation 28 (3) 106-114, 2014.

28

THE HELIUM QUESTION

DURING THE DECAY of various elements' radioactive isotopes, helium nuclei are given off, and this process is part of a dating technique utilized to come up with long ages (many millions of years). But there is bias, misunderstanding, and misinterpretation, as many Christian scientists point out. Richard and Tina Kleiss sum up the helium action in the following easy to understand manner. As radioactive elements decay, helium atoms are deposited into rock crystals. Helium atoms are so small that they easily leak through the rock structure and escape into the atmosphere. The rate at which helium leaks out of the rocks has recently been measured, and all of the helium produced by radioactive decay should have left the rocks if these formations were millions of years old. Yet the rock still contains much of the helium produced by this decay process. **This helium is still locked into the rocks because there hasn't been enough time for it to escape from the rocks.** This means that the radioactive decay could not have happened billions, or even just millions of years ago. Apparently there was a burst of radioactive decay within the last 10,000 years, possibly corresponding to when the earth was created, during the worldwide flood, or both. As the Kleiss team points out, "these are the types of scientific observations which make creation science so exciting."[125]

For those interested in studying this very important helium diffusion issue, which is **strong** evidence for a young earth, at a deeper and more detailed scientific level, check out the R.A.T.E. study-*Radioisotopes and the Age of the Earth*-by the scientists at the Institute for Creation Research.

---

[125] Richard and Tina Kleiss, *A Closer Look at the Evidence*, Search for the Truth Publications, 2017, April 3, Physics- daily devotional

PART III

IS CULTURE RELIGION EXTERNALIZED?

29

THE PROBLEM WITH ASSUMPTIONS

AS STRESSED ALREADY, thoughts are *things*. Most actions-good or bad- begin in the mind as a thought. Most of the time a beautiful painting, a great golf shot, an inspirational symphony, an awesome skyscraper, or a terrible crime all began as a thought or idea. There is much truth to the maxim, "As a man thinketh in his heart, so is he" (Proverbs 23:7). Foundational worldview thoughts are those that form the platform, basis, or filtering grid through which one views the world, and how one interprets not only the physical make-up and environment we live in, but possible *purpose*, and the flow of events, time, and life on planet earth, and even beyond. These might be called presuppositions, or assumptions, and are powerful motivators in our lives that often influence our decisions, lifestyle and how we spend our time and money more than most of us even realize. To a great degree they help establish our values and how we think, feel, and act.

As a college student reading Hemingway's *Islands in the Stream*, I was struck by the words, "a man has to live inside himself wherever he is." It just resonated with me at the time; it was not something I had previously considered or even thought much about. The inner life of the mind is indeed unique and powerful, but needs tended and cultivated. Even scripture says "be transformed by the renewing of your mind" (Rom. 12:2). The mainline core beliefs, or **foundational assumptions** (*starting presuppositions*, if you prefer), can even determine one's interpretation of history or expectation for the future, or the presence of innervating *meaning* and *hope* . . . or the lack thereof.

What if the bedrock assumption, the starting presupposition, or the worldview foundation platform is *false*—or *wrong*? Ralph Waldo Emerson, the confused nineteenth-century Unitarian turned Transcendentalist, developed the belief—or foundational assumption—that God could best be discovered by deeply looking inward into one's *own* self, and that nothing can truly bring you peace but yourself, and the triumph of your *own* principles. He was definitely an influential dude in the mid-1800s, and a forerunner of today's *relative* morality and psychology's self-esteem theory. Well, let's just pause and reflect for a moment, and consider a different take. The Bible says that the peace one gets as a follower of Jesus

Christ "passes all understanding"; history and untold numbers of miraculously changed lives (including my own) support *this* claim, which is in opposition to Emerson's worldview. Wrong starting points or assumptions can not only lead to confusion and troubled personal relationships, but misinterpretation of events, history and the physical world, as well as poor government policy and a myriad of other problems, and can sometimes be downright dangerous with a horrific cascade of negative consequences.

The all-important creation vs. evolution origins debate is a fascinating study of this question, what I think of as *the assumption problem*. Think about this issue calmly and objectively for a moment, from this angle:

> We all have and live in the same world. We all have the same mathematical formulas and scientific laws. All of us have the same geological strata, canyons, mountain ranges, giant river valleys, volcanoes, coal seams, dating methods, laboratory equipment and techniques, access to telescope and space probe data, computer access, animals and plants and microbes, fossils, etc. to study, observe and analyze. But somehow we interpret the findings and observations in *very* different ways. How can this be? All of us involved or interested in the debate and the question of where we came from have the *same data and laws of science*, as mentioned, and it is not really just a science vs religion debate. So, what gives? The answer: a profound difference in our foundational starting points—our **starting assumptions**.

The primary assumption of modern science for the last 150 years is naturalism, or evolutionary materialism—that things somehow made themselves (including the stunningly complex DNA code)—and that they developed from **inorganic non-living chemicals**. Part and parcel of that uniformitarian deep-time worldview includes no input—or need for—a Creator God . . . truly just a "material" world, as Madonna used to claim (I wonder if she's still thinking about taking out President Trump, and blowing up the White House?). Geneticist, human genome sleuth and "gene gun" inventor Dr. John Sanford, in his classic book *Genetic Entropy*, describes it a bit differently as he evaluates and describes this *un*scientific starting assumption of long-ages Darwinian evolution, as built upon *the Primary Axiom* "that man is merely the product of *random mutations* plus *natural selection*,"[126] as **totally unsupported and unjustified**. This para-

---

[126] Ibid, Sanford, *Genetic . . .* p. v.

digm-shaking book, on the stunningly surprising discovery of the quick-ly-increasing genetic load (piling-up of deleterious mutations) in the human genome, is a significant exposure of the impossibility of this worldview—or starting assumption—based on real biochemical and genetic evidence.

The alternative, radically different, *opposing* starting foundation (or assumption), for uncountable numbers of Jews and Christians throughout the ages, and millions more today including Bible-believing young-earth creation scientists: trust in the veracity of the Bible and the historical record of Genesis, with recent creation (a little more than 6,000 years ago) by an infinite personal Creator God, and an early history of catastrophism (not uniformitarianism) with an earth-reforming global worldwide Flood (some 1,750 years after Creation, or some 4,300 years ago). This would have been followed by *one* resultant signifi-cant Ice Age, along with the Confusion of Tongues at Babel. It is hard to imagine two starting assumptions and worldviews being more different . . . because— it's worth mentioning again —**we've got the *same* data**. What gives?

In their wonderful book, *The Fossil Record*, Dr. John Morris and Frank Sherwin touch on the problem of opposing starting points:

> The study of data must inevitably begin with a set of assumptions, which dictate which data are deemed important, which measurements are chosen, and which experiments are run. The interpretations follow, but **the assumptions**, or worldview, held at the start are easily *the most important part* of the interpretation process.[127]

Are you starting to sense the influence of entrenched worldviews, as well as outright bias, with the unavoidable subsequent suppression of crit-ical thinking, even in science? It is unfortunate, but is especially rampant in historical science, geology, paleontology, biology and the *origins* dis-cussion. Consider for a moment the possibility of truth to the claim of Dr. Sanford (and a quickly growing number of thousands of other scien-tists and researchers), who agree that the so-called *Primary Axiom*—man being merely a product of random mutations through natural selection—is untenable and unsupportable by the evidence.

The historical consequences and ramifications are profound when one looks at well over a century of false teachings and programs. We're talking about countless misled and confused students that were taught not only false history and science, but a purposeless, no-hope worldview; wasted

---

[127] Ibid, Morris and Sherwin, p. 29.

millions and billions in grants and government programs; and fruitless expenditures on meaningless lines of evolution research. As evolutionary materialism led the elites (and then some thugs, like Stalin) of Western Civilization into the 20th century, it turned into the bloodiest century in recent millennia, or what Dr. Carson Holloway called "organized mass bureaucratic slaughter."[128] Let's look at a few of these culture-changing actors and villains a bit closer, who operated from **a false foundational assumption**. It is very important relative to the *how we got in the mess we're in* question, as well as our young people knowing truth, and living as grounded real men and women, and contributing cultural "salt and light" to the culture.

- Mao Tse-tung (now often Mao Zedong) of China: Many of the evolutionary writings and arguments of Charles Darwin's champion "bulldog" Thomas Huxley were already translated into Chinese before 1900 and quickly became widely read and embraced, as the Chinese already generally believed in the origin of the universe by natural processes. Far east expert and medical historian Dr. Ilza Veith, sixty years ago in 1959—the year of the Darwinian Centennial—said this: "But it was Darwinism, speaking through Huxley, and made to appear organically related to ancient Chinese thought on evolution, that furnished the intellectual basis for China's great upheaval beginning in 1911."[129] For those of you who were not taught this in your probable revisionist history classes, this Darwinism-inspired upheaval and metamorphosis led to the ruthless dictatorial leadership of Mao Tse-tung and the starvation and slaughter of millions in 20th century China, as Marxist communism spread forth. It is important to remember the primary tenet (or first plank) of communism: *there is no God*. It is still fascinating to me, reading the comments of a respected Chinese scholar, relative to the catastrophe of WW1, even back in 1920:

This great European war has nearly wiped out human civilization; although its causes were very many, it must be said that the Darwinian Theory had a *very great* influence. Even in

[128] Dr. Carson Holloway, *Darwin's Deadly Legacy-The Chilling Impact of Darwin's Theory of Evolution*, (DVD) hosted by D. James Kennedy, PhD, Coral Ridge Ministries, Fort Lauderdale, FL, 2006

[129] Ilza Veith, "Creation and Evolution in the Far East," in *Issues in Evolution*, Sol Tax, ed. (University of Chicago Press, 1960), p.16

China in recent years, where throughout a whole country men struggle for power, grasp for grain, and seem to have gone crazy, although they understand nothing of scholarship, yet the things they say to shield themselves from condemnations are regularly drawn from Yen Fu's translation of T. H. Huxley's *Principles of Evolution*. One can see that the influence of the theory on man's minds is **enormous.**[130]

– Communism Founders and bullies Marx, Engels, Lenin, and Stalin: Atheistic and vehemently anti-Christian communism founders Karl Marx and Friedrich Engels received virtual shots of adrenaline when Darwin's 1859 *Origins* book came out, as it supposedly supplied the scientific and biological justification for their own non-God material-istic idea, and they felt it could be the death knell for Christianity. Marx was so enamored he wanted to dedicate part of his own magnum opus, *Das Kapital,* to Darwin, but it seems that idea was suppressed by Darwin's wife. Vladimir Lenin embraced the writings of Karl Marx and applied Marx's communism idea in founding the Russian Communist Party, initiated the Bolshevik Revolution in 1917, and subsequently developed the Soviet state, while contributing to the starvation and deaths of millions. Josef Stalin took over after Lenin's death, and with brutal iron-fisted rule held sway for nearly 30 years in the Soviet Union, with millions more starved to death, executed or exiled to the Siberian work camps and gulags; the frustrated former altar boy cast his remnant Christian belief down the drain **after reading about Charles Darwin's evolutionary idea.** These four men and their henchmen and followers are indirectly or directly responsible for many millions of deaths in the last hundred years (the estimates vary from 50 to over 100 million), and unimaginable human suffering.

– Francis Galton: Charles Darwin's cousin who embraced the evolution idea and developed and promoted his new *pseudoscience* of eugenics—essentially applied Darwinism—as he became mesmerized by Herbert Spencer's "survival of the fittest" concept. He wanted to **speed up and guide** the supposed evolutionary process and the expected improvement of the human race—essentially playing God—through various practices and policies, liked forced sterilization, with his "scientific" racism.

---

[130] Originally reported by Ssu-yii and John K. Fairbank in *China's Response to the West* (Cambridge: Harvard University Press, 1954), p. 267. Cited in I. Veith "Creation and Evolution," pp. 16-17.

- Margaret Sanger: Founder and promoter of the 20th century abortion industry and Planned Parenthood, which is now a multi-billion dollar industry operating in over a hundred nations, after its nefarious beginnings in a run-down Brooklyn neighborhood prior to WWII. Sanger was a follower and disciple of Thomas Malthus and his radical **racist population control** idea, Galton's pseudo-science of eugenics, and Darwin's evolutionary idea and was one of the early catalysts for the 20th century sexual revolution. Like her heroes, mentors and many of her lovers (which included George Bernard Shaw, H. G. Wells, English writer Arnold Bennett and Havelock Ellis) she despised Christianity and **wanted to speed up the human evolutionary process**. Dr. George Grant describes her worldview:

> She was thoroughly convinced that the "inferior races" were in fact "human weeds" and a "menace to civilization" . . . and had come to regard organized charity to ethnic minorities and the poor as a "symptom of malignant social disease" because it encouraged the prolificacy of those "defectives, delinquents, and dependents" she so obviously abhorred. She yearned for the end of the Christian "reign of benevolence" . . . [and] her greatest aspiration was "to create a race of thoroughbreds" by encouraging more children from the fit, and less from the unfit.[131]

- Adolph Hitler: German Chancellor/dictator, head of the infamous Nazi party, instigator of the bloodiest war in history (WWII), architect of the Holocaust with at least six million Jews killed in the Nazi concentration camps, along with many more gypsies, Slavs and others all because of his perverted desire **to speed up the evolutionary process** and purify the Aryan race. His diminutive and devoted Reich Minister of Propaganda, Joseph Goebbels, classified such people groups as merely "human eaters." *Selection*—one of Charles Darwin's favorite terms—was used and implemented constantly by the Nazis, and not just against the Jews in the concentration camps. Thus this **applied Darwinism**, or targeted *artificial* selection, not only involved them and other ethnic undesirables, but uncountable numbers of the chronically ill, the mentally compromised, and the permanently maimed or disabled. These unfortunates were *selected* also, and herded to the new euthanasia centers beginning in the late 1930s. This Social Darwinism, German militarism, and racism reached their

---

[131] George Grant, *Killer Angel*, Standfast Books & Press, Franklin, TN, 2014, p. 50

historical peak under the evolution-driven Fuhrer. If you doubt Hitler's enthusiastic evolutionary mindset, just look at his words and read his book, *Mein Kamph* (*My Struggle*). His words: "the purity of the *racial* blood should be guarded, so that the best types of human beings may be preserved." One of Hitler's worst henchmen and most devoted followers was SS officer Reinhard Heydrich, known as *The Hangman*, as well as *the Butcher of Prague*. Heydrich played a key role working with Heinrich Himmler in organizing the *Holocaust* around the beginnng of WWII. As these Nazis prepared to launch their heinous race purification/**Social Darwinism program**, and the murder of millions of Jews and others, Heydrich uttered his now famous prediction: "Darwin would be astounded at the progress we're going to make in one year."

After stunning unimaginable human suffering, and multiple tens upon tens of millions of deaths, there was finally somewhat of an international elitist reaction by some of the materialistic social Darwinists against the philosophy and worldview undergirding the Nazi mindset after their defeat. This fortunately included, albeit reluctantly in some quarters, opposition against the **racist** idea of different intellectually qualified (evolved) people groups (sort of an evolution-based, contrived human caste system). In the late 1940s after WWII, students began to be taught about the evils of Hitler and the Nazis, and many of the evolution-driven eugenicists and academics backed off from their enthusiastic endorsement of ethnic cleansing attempts, but the mid-twentieth century educational elites—with their secular materialistic worldview—were lukewarm (some still are), and didn't tell the whole truth. Dr. Henry Morris decades ago described Hitler and the post-war academic mindset well: "The evolutionary philosophy that had energized them . . . [was] still alive and well. In fact, whereas every public school student is now well instructed in the evils of National Socialism (Nazism), they are almost *never* taught that it was **founded on evolutionism**. This has been an amazing cover-up, **even a rewriting of history**. Modern evolutionists react angrily when reminded that *evolution provided the rationale for Nazism*, but it is true nonetheless."[132] Dr. Morris follows with this: "Sir Arthur Keith, the leading British evolutionary anthropologist of the first half of the twentieth century, wrote a remarkable book right after World War II, titled *Evolution and Ethics*. Having endured with other Londoners the terrible bombing of Britain by the Hitler's Luftwaffe, Keith certainly had no affectation for Hitler. Nevertheless, in consistency with his own evolutionary commitments, he honored Hitler as a thoroughgoing evolutionist, in practice as well as theory":

---

[132] Dr. Henry Morris, *The Long War Against God*, Master Books, Green Forest, AR, 2000, pp. 75-76

> To see evolutionary measures and tribal morality being applied rigorously to the affairs of a great modern nation, we must turn again to Germany of 1942. We see Hitler devoutly convinced that evolution produces the only real basis for a national policy. . . The German Fuhrer, as I have consistently maintained, **is an evolutionist; he has consciously sought to make the practices of Germany conform to the theory of evolution.**

That particular citation is taken from Keith's 1947 book (p. 230), where Sir Arthur Keith, a hard-core evolutionist himself, in trying to explain human criminal behavior and the brutal social Darwinism of Nazi Germany, also says this: ". . . as we have just seen, the ways of national evolution, both in the past and in the present, are cruel, brutal, ruthless and without mercy. . . . **the law of Christ** *is incompatible with* **the law of evolution.**"[133] Do not misunderstand; the internationally-known Keith laughed at the "law of Christ," was an atheist and racist evolutionist, and supporter of social Darwinism. (Does anyone besides me ever grow weary of all the British "Sirs"?) Are you readers puzzled, or perhaps downright ticked off, that you were **never taught** that a critical component of the driving philosophical and worldview force behind Marxism-communism-socialism, Lenin and Stalin, brutal Chinese Chairman Mao, Hitler and the Nazis , and the unfolding of the bloodiest century in world history, was Darwinian evolution? This includes its iconic principles such as: "kill or be killed," "only the strongest survive," "survival of the fittest" and "the end justifies the means." Is it not stunning, and downright tragic, how certain people in history want to play God?

Those starting presuppositions, or foundational assumptions, are *always* important . . . **even more so *when they're wrong*.**

---

[133] Arthur Keith, *Evolution and Ethics*, Putnam, New York, 1947, p. 15.

30

THOMAS MALTHUS, MARGARET SANGER AND PREMEDITATED ETHNIC ANNIHILATION

SOME 200 YEARS ago, Englishman Thomas Malthus, with his ungrounded theory of exponential population growth, convinced an entire generation of intellectuals and industrialists, professors and academics, scientists and societal pundits, and most western politicians that the world was headed at increasing speed towards a potentially catastrophic population explosion, with many serious collateral problems. His views came to dominate social policy in Europe and America in the 19th century and well into the 20th, and contributed mightily to the "cultural mess we're in." I was even taught—more lightly and with less imminent emergency—the same drivel as a college undergraduate student in the 1970s, albeit with the racism, elitism, and some of the severe control methods minimized or eliminated. Malthus' solution and policy recommendations *did not* include humanitarian aid, Christian charity and outreaches, philanthropic gifts and assistance, or government support for the poor, the sick, the mentally compromised, or the downtrodden.

On the contrary, Malthus felt the best policy to avoid this fast-approaching omnipresent population problem (real only in his mind), was to go *opposite* the Christian ethic, by *exacerbating and accelerating* their decline, starvation, and illnesses, with the compromised lower class masses' eventual elimination. Instead of *helping* the poor, Malthus frowned on any charity, training, hygiene measures, etc., and said this: "We should facilitate . . . the operations of nature **in producing this mortality** . . . and sedulously encourage other forms of destruction . . . Instead of recommending cleanliness to the poor, we *should encourage contrary habits*. In our towns we should make the streets narrower, crowd more people into the houses, and **court the return of the plague** . . . build our villages near stagnant pools . . . But above all, **reprobate [reject] specific remedies for ravaging diseases**; and restrain those benevolent, but much mistaken men, who thought they were doing a service to mankind by projecting schemes for the total extirpation of particular disorders."[134]

This may be stunning to modern ears, but these dire predictions and proposals of Malthus, combined with a lightning-fast acceptance of Darwinian evolution after the American Civil War, fueled not only an acceleration of racism, social Darwinism, and the widespread practice of Malthusian-based eugenics (to speed up the evolutionary process), but was the catalyst and justification for atheist Margaret Sanger's first seedy abortion clinics. These eventually led to Planned Parenthood, and her initial goal of ethnic annihilation was off and running. She was determined to do her part to exterminate what she and the materialistic, **evolution-driven** intellectual elites considered un-favored or inferior races—particularly non-whites. This elitist group's shared core belief—or worldview—a century ago: non-belief in God, total acceptance and application of Darwinian evolution and "survival of the fittest," relative morality, and some form of **purification of the human race**.

Incidentally and ironically, a recent report by one of my least favorite organizations—the United Nations—called *World Population Prospect 2019: Highlights*, now actually allays all fears of global overpopulation, in case any remnant Malthusian fans are still fretting over all the people (did you see that, [former Stanford Professor] Paul Ehrlich?). It was probably never pointed out to most of you readers in your censored scholastic career that ardent racist Charles Darwin was one of the key philosophical heroes to Ms. Sanger and her ilk. Significantly, the very sub-title of Darwin's world-changing *On the Origin of the Species* magnum opus, first published in 1859, is "Or the Preservation of Favored Races in the Struggle for Life."

Favored races? What does Darwin mean by *favored*, or even *race*, for that matter? The frustrated former divinity student surely didn't get it from the Bible; it clearly says "God is no respecter of persons" (no preferential favoritism), and *never* **mentions** *race*—just tribes and people groups. What is he talking about? The refined, well-bred Englishman later explains himself; he's talking mainly about what he considers *inferior* human beings, specifically *non-white* human beings. It is quite clear how he feels about people of color, race and racism in his follow-up book, *The Descent of Man*:

> At some time period, not very distant as measured by centuries, the civilized races of man will almost certainly exterminate, and replace, the savage races throughout the world. At

---

[134] Allan Chase, *The Legacy of Malthus: The Social Costs of the New Scientific Racism*, Knoph-Doubleday, New York, 1977, p. 7

the same time the anthropomorphous apes, as Professor Schaaffhausen has remarked, will no doubt be exterminated. The break between man and his nearest allies will then be wider, for it will intervene between man in a more civilized state, as we may hope, even than the Caucasian, and some ape as low as a baboon, instead of as now between the Negro or Australian and the gorilla.[135]

To many of us moderns, especially those of Christian persuasion, this almost unbelievable racist attitude and worldview by the genteel Englishman who never had to work for income to support himself or his family, is simply stunning. Yet his evolution idea—really *just a story* (which still undergirds our entire education and social system)—is a worldview which was passed on and embraced by people of academia, industry and government and contributed to suffering and death for untold millions through aggressive European colonialism (especially in Africa and Australia); Marxism/communism (through Marx, Lenin, Stalin, Mao Zedong, and Fidel Castro); and Herbert Spencer's Social Darwinism—taken to a tragic extreme by Adolph Hitler—and also through prominent European and American business and industrial leaders. It was also the driving force behind the promotion and practice of Francis Galton's *eugenics* idea and then abortion—the latter spearheaded by American Margaret Sanger. For the full true story of Ms. Sanger, the biography *Killer Angel* by Dr. George Grant is highly recommended.

It deserves mention here that Ms. Sanger considered the "inferior races" (Slavs, gypsies, Jews, non-whites, and others) essentially scum and a detriment or "menace" to civilization. Like her philosophical mentor Malthus, she despised assistance, charity and Christian support for the poor as this only prolonged their existence; one of her main goals was to rid the world of such weak, lesser-stock humans and create a "race of thoroughbreds."[136] Thus came forth the first back alley Margaret Sanger abortion "clinics" (1920s) of the American Birth Control League, the forerunner of Planned Parenthood. Subsequent abortion locations were always in the poorer, more run-down parts of big cities where it would be easier to attract these undesirables and curtail their reproduction, which

---

[135] Charles Darwin, *The Descent of Man*, 2nd edition, John Murray, London, 1882, p. 183.

[136] Margaret Sanger, *The Pivot of Civilization*, New York: Brentano's, 1922, pp. 23, 107

hopefully would eventually lead to their total demise and extermination (wow, talk about secular long-term thinking). All non-Aryans were eventually targeted; she referred to them as "dysgenic races" or groups. This soon included Hispanics and Catholics. Relative to her "Negro Plan" proposal for the South, she said this: "[they] still breed carelessly and disastrously, with the result that the increase among Negroes, even more than among Whites, is from that portion of the population least intelligent and fit."[137]

Such scathing, racist thinking and words, and I think often about the virtual uncountable number of abortions performed on African-American women by the American Birth Control League and then Planned Parenthood (what an inappropriate name!) the last 90-plus years. It is worth repeating, with emphasis: the founder Margaret Sanger's original goal was to eventually wipe out all people of color, strongly undergirded by Thomas Malthus' population control philosophy, the Eugenics movement, Darwinian "survival of the fittest" evolutionary thought, and orchestrated partially through her abortion vehicle. And the American government regrettably still partially finances and subsidizes her heinous company—Planned Parenthood.

It should be mentioned that the Eugenics movement spawned in the mid-late 1800s from Charles Darwin's evolution idea, and was initiated and boldly promoted by his first cousin, Francis Galton. He grasped the idea of survival of the "fittest," and looked at human heredity partly from a statistical, empirical or mathematical perspective. Galton was very influential, and he came up with the term *eugenics*, from the Greek meaning 'good in birth' as well as 'noble in heredity,' and his goal was to suppress the poor, the indigents, certain ethnics and various other "undesirables." His goal: to improve and even purify the human race, by "giving the more suitable races or strains of blood a better chance of prevailing over the less suitable"[138] . . . just amazing, and downright scary.

Abortion, euthanasia and now stunningly *even infanticide* in these United States—spearheaded and promoted by Democrat Party leaders like New York Governor Andrew Cuomo and Virginia Governor Ralph Northam (unbelievably, a pediatric neurologist by former occupation)—are unsurprising consequences of a culture that has diminished and denied the Creator God, and *His image* in the eternal soul of man. How many murders of obviously alive, fully-formed, uniquely programmed, divine-image-bearing defenseless human beings, through the horrific practice of abortion, have been committed *legally* in the United States since the 5-4

---

[137] Linda Gordon, *Woman's Body, Woman's Right*, Penguin, New York, 1974, p. 332

[138] Otto Scott, "Playing God," *Chalcedon Report*, no. 247, Feb., 1986: p. 1

Roe vs Wade Supreme Court vote in 1973—with a big chunk of them by Planned Parenthood? The answer: Over 50 million (some claim as high as 54 million) through mid-2019. I remember as a young self-centered college student in the 1970s, thinking that those highly educated Supreme Court dudes in the black robes were surely smarter than I was, and therefore I figured I was probably okay with their decision. Only when I accepted Jesus Christ as my Creator, Savior and Redeemer several years later did my stance do an about-face.

Significantly, evolutionary diagrams and arguments were presented in 1973 to those God-like Supreme Court justices, as supporting evidence of the view that the human embryo and fetus went through an early developmental pattern, or sequence, that mimicked or hearkened back to man's supposed long ago pre-human evolutionary stages. These were made famous by unscrupulous German evolutionist Ernst Haeckel's fraudulently doctored embryonic drawings, known to be fake *even then* by biologists who were Christians, but still somehow allowed to be presented in support of legalizing abortion to the high court. The message from the falsified embryonic diagrams by the abortion advocates: just look at the pictures; surely any intelligent person can see that in the early stages in the mother's womb, it's just a small gelatinous mass, or an amphibian-like or perhaps a semi-reptilian ancestral form—*not really human yet*—and therefore it's okay to kill the inconvenient little blob and throw it in the trash can.

This false, made-up idea became famously known as *ontogeny recapitulates phylogeny.* Indeed, it has an impressive scientific ring to it. **It is simply a bunch of tragic hogwash**. Amazingly, there are some biology textbooks that *still include* the fake pictures of "Haeckel's Embryos." Ernst Haeckel was a philosophical hero to not only Adolph Hitler but also to Sigmund Freud, and German philosopher Friedrich Nietzsche, who was impacted in a powerful way by the Darwinian evolution idea and its superior human races theme. This led to his famous "God is dead" worldview proclamation and his claim that "theism is the refuge of *weak* minds." This was tragically ironic—considering his own mind challenges—with his descent into insanity before dying of pneumonia in 1900 at age 55. Nietzsche energetically promoted the eventual evolution of the "master-class" into "ubermenschen," or *supermen*, who would take power and bring humanity to the next stage of its **evolution**, unencumbered by any religious or social mores.[139] The young philologist-turned-philosopher was very popular with European elites and championed warfare, colonialism and eugenics to get rid of the unfit and inferior races in accelerating the purification of the

---

[139] John MacArthur, *The Battle for the Beginning*, W Publishing Group (Div. of Thomas Nelson, Inc.), 2001, p. 16

human race. Always beware of these elitists and experts. Nietzsche was also the pundit who said, "Crooked is the path of eternity." Huh? What? As either of my sons would likely have responded, even in elementary school, "Dude, how do you know?!"

The imperialism, war, and aggressive racist colonialism by the European powers in the late nineteenth century were driven by the evolutionary worldview. Just some thirty years after Darwin's famous *Origins* book was published, the following was part of a commentary that was widely read in the United States: "The greatest authority of all the advocates of war is Darwin. Since the theory of evolution has been promulgated, they can cover their natural barbarism with the name of Darwin and proclaim the sanguinary instincts of their inmost hearts as the last word of science."[140]

I am hopeful that you astute readers realize more than ever that *thoughts are things* and that the so-called **origins debate** *is not a side issue*. Think soberly about this: by a 5-4 vote of **unelected** judicial officials, the Supreme Court of the United States of America—formerly the land of the free and the *safe* (as well as the brave)—legalized infant human execution based partially **on fraudulent evolutionary arguments and pictures** (purposely designed to deceive) . . . and here we are, with an American legacy of 50+ million murdered babies, and probably many more to come. It has been known for decades that babies, even in the first trimester, experience extreme discomfort and pain in the lethal abortion procedure. **Everyone** should take 29 minutes of their life to watch the 1984 documentary film, *The Silent Scream*, narrated by a *former* abortion provider. Psalm 139:13-16 says the Creator-God "covered me in my mother's womb," which means shielded and protected, and [the baby] is "fearfully and wonderfully made," where wonderfully essentially means *differently* and uniquely. This is followed by the claim of being "made in secret", and "curiously wrought," which interestingly means *embroidered*, a stunning description of the profound double-helix DNA master molecule which directs the differentiation and development of every miracle infant. You young and millennial men and women out there need to know this stuff. . . as well as the extreme importance of thoughts and ideas, and **the danger of false assumptions**.

This disregard for human decency, mocking and disbelief of "made in the image of God," along with misguided embracement of Malthusian population control, eugenics, and policies to speed up the supposed **evolutionary purification process** of mankind, contrasts sharply with the Christian worldview. Incredibly, these evolutionary-driven eugenics

---

[140] Max Nordau, "The Philosophy and Morals of War," *North American Review*, 1889, #169, p. 794; cited from Richard Hofstadter's *Social Darwinism in American Thought* (1944), p. 171

policies were enthusiastically endorsed by intellectual elites, industrialists, and our country's leading corporate philanthropies in the late 1800s and first half of the 20th century. The Carnegie Institution played a key role, as did the Rockefeller Foundation. In fact, some historians and scientists believe the Rockefeller Foundation's large financial grants to German scientists and researchers contributed to and indirectly led to Josef Mengele's horrific experiments at Auschwitz, and the extremes of German social Darwinism.[141] This racist philosophy quickly spread to the classroom. I can even remember a discussion with my paternal grandfather in the early 1970s, where he described being taught this in high school (he was born in 1906). Do you understand the point here? **A *superior* white race—flagrant racism—was taught in some American schools** in the first half of the 20th century.

British, American, German and other scientific elitists even hustled to gather museum and zoo specimens and study heads of the supposed "sub-humans" for display and analysis, particularly those of Pygmy and Australian Aboriginal stock. I'll never forget my stunned disbelief when I first learned the sad story of the Pygmy man Ota Benga from the Belgian Congo. After his family was murdered and mutilated, Ota was brought to the United States in 1904 for zoo display, initially at the 1904 St. Louis World's Fair, while falsely being touted as a mentally compromised probable missing link. He was later placed in the Bronx Zoological Gardens and then the New York Zoological Park (the largest zoo in the world then) in the *monkey house*, quite often with an orangutan, to the delight of thousands of New Yorkers every day. This sad and stunning racist display, of course, insinuated that blacks were an inferior race. A protest response eventually came forth from some sharp and savvy, local African-American ministers with admirable intestinal fortitude:

The exhibition evidently aims to be a demonstration of the Darwinian theory of evolution. The Darwinian theory is *absolutely opposed to* Christianity, and a public demonstration in its favor should not be permitted.[142]

---

[141] Edwin Black, *War Against the Weak, Eugenics and America's Campaign to Create a Master Race*, Four Walls Eight Windows, New York, 2003, pp. 93-95, 243-245, 258, 283-285, 288, 294-298, 302-303, 306-308, 313-314, 349, 364-365, 369-370, 419-420, 422

[142] P. V. Bradford and H. Blume, *Ota Benga; The Pygmy in the Zoo,* St. Martin's Press, New York, 1992, p. 183

I agree, but those days are long gone, as ungrounded displays and demonstrations in favor of evolution are the *only* ones permitted. Today's big print-media bastion of anti-Christian secular groupthink and evolutionary humanism—*The New York Times*—responded at that time in typical fashion to the concerned black ministers:

> One reverend colored brother objects to the curious exhibition on the grounds that it is an impious effort to lend credibility to Darwin's dreadful theories . . . the reverend colored brother should be told that evolution . . . is now taught in the textbooks of all the schools and it is no more debatable than the multiplication table.[143]

Wow! That supercilious highbrow attitude sounds like something I heard Bill Maher say on TV a few years ago as he was slamming orthodox Christians and creation believers. Or perhaps Garry Wills, who some fifteen years ago in a critical editorial of born-again Christian President George W. Bush in the *New York Times*, frustratingly admitted "we don't get Christianity." Back to the earlier condescending *Times* response to the black pastors a century ago, it actually **is very debatable**; it's just hard to do because **it is *not permitted* in public schools (bias and censorship), or even most universities**. How would you accurately describe the situation in Big Mainstream Media and many American colleges and universities today? *Cultural Marxism* would probably be the most succinct term, and it is flat-out inimical to the American ethos. Many parents have little idea what's really going on, and what they're paying for.

There is even documented evidence that the remains of some 10,000 of the Aboriginal people were shipped to the British museums in a frenzied attempt to prove the widespread belief that they were the "missing link." The rarified air of the Smithsonian Institution—for the same reason—is actually polluted too, as they hold the remains of 15,000 individuals of various races.[144] To this day, the ungrounded long-ages/evolution foundational assumption dominates there. I always wondered how the talented and charming Australian Aborigine -Evonne Goolagong Cawley -my

---

[143] "Topics of the Times; The Pygmy Is Not the Point," *New York Times*, Sept. 12, 1906, p.8

[144] Ken Ham, Carl Wieland, Don Batten, *One Blood-The Biblical Answer to Racism*, Master Books, Green Forest, AR, 2002, pp. 119-120

favorite female tennis player back in the day, who won 14 Grand Slam titles, felt about all of this. She grew up as a young girl in 1950s and early 60s Australia where there was a fairly strong evolutionary worldview that the dark-skinned Aborigines were a *less evolved* "race." Just some 45 years before her birth, Australian aborigines were listed in a Sydney museum booklet as an animal. Is it any wonder that we still have a racism problem today?!

Racism has existed for centuries, but as the oft-quoted evolutionist Stephen J. Gould alluded to on several occasions, it *increased by orders of magnitude* after Darwin's Origins book came out (1859) and was quickly adopted by academic and intellectual elites. There are many factors contributing to this, but this dark chapter just discussed is one of the important ones, which has been stifled, successfully hidden and suppressed by various modern academic elites, many of whom are the *progeny*—so to speak—of the early 20th century evolutionary humanists. One could even make a solid argument that this pseudoscientific race-based movement, specifically catalyzed in the United States in 1904 by political, business, and academic elites, had major influence a short time later on Adolph Hitler and the German Nazi Party.[145]

This contrast of worldviews and cultural fruit between evolutionary materialism and true Christianity with an infinite Creator-God is profound, exhibited particularly in the unique Christian pillars of love and the worth— or value—of *every* human being. This principle was apparent even nearly 2,000 years ago to Aristides, as he described Christians to the Roman Emperor Hadrian:

> Oh, they love one another. They never fail to help widows; they save orphans from those who would hurt them. If they have something, they give freely to the man who has nothing; if they see a stranger, they take him home. And Hadrian, they are so happy! It must be their love.

---

[145] Ibid, Black, pp. xv-xxv, and 3-371

31

HOMOSEXUALITY, GENDER DYSPHORIA, MAYOR PETE . . . AND THOSE RADICAL FEMINISTS

IT SHOULD BE increasingly obvious that one of the main issues discussed in this book is how we arrived at the divisive, dysphoric, politically correct, *anything goes* cultural maze we're in. Anybody that denies that we're in a mess, or dealing with unprecedented cultural confusion and division while drifting further away from our "ancient landmarks" on a sea of relativity, is simply not in the ballgame and likely couldn't care less. The suppression and denial of America's Judeo-Christian Founding—based on a Creator God and biblical precepts (including the *fallenness* of man)—while naturalism (or evolutionary materialism) has become the accepted and often celebrated **replacement foundational assumption** or starting point, is irrefutably at the heart of the problem. For some time now, the resultant and predictable operative principle and mindset is *relative morality*, with everyone doin' their own thing, setting their own rules, including those relative to sexual behavior, attitudes, and preferences. This worldview has spawned a myriad of problems and bad fruit, far too varied and numerous to all be covered in this book. A few significant ones in addition to the tragic results of Social Darwinism, include far more deaths worldwide from AIDS than all American war casualties in history, well over 54 million murdered infant Americans through abortion, unprecedented numbers of mass shootings by young men in the last two decades, an explosion in divorce and drug use, sexual immorality, rising suicide rates, as well as the already mentioned attack on real manhood and family values.

The evidence seems to now overwhelmingly support the contention of political scientist Stephen Baskerville that in the last half-century, the most dominant manifestation of this metamorphosis and worldview shift is "the emergence of a political agenda and ideology that derives political power

from demands to control and change the terms of sexuality."[146] Sexual revolutionaries, radical feminists, and assorted leftist university professors and
Women's Studies advocates do not even deny this; in fact, they sometimes
boast about it. Back to the beginning, though, or **foundational assumption:** if on a *truth search*—surely a worthy and honorable quest for people
of intelligence and character everywhere—it begs the question: What if
the starting foundational assumption, or *primary axiom,*[147] is wrong?

After objectively examining real facts and cultural fruit, real history,
and particularly *real science*, a major contention of this book is that the
modern (and *postmodern)* humanistic foundational starting assumption—
Darwinian evolution, or naturalism—is *unjustified and false.* As a former
undergraduate biology major (and evolution believer), it undergirded
nearly every class I took—including many *non*-science classes. Objective
classroom discussions of *opposing* ideas and evidence, like the evidence
for creation, intelligent design, or a recent global flood—with critical
thinking application—were simply not tolerated. If evolution is true, what
are the professors and white lab coats *afraid of*? It reminds me of the
mellifluous-sounding *National Center for Science Education*, an organization that promotes the teaching of evolution-and *only* evolution-as the
explanation for origins, in the public schools and museums (and anywhere
else they can influence) across America, with the ever-present support,
when needed, of the American Civil Liberties Union (ACLU). I remember a
number of years ago when I discovered that its atheist director, Dr. Eugenie
Scott, in apparent frustration, issued a directive to teachers, professors,
and assorted minions to **quit debating creation scientists** on college campuses, as they (the creationists) seemed to *always win* the debates.

With further reflection on the widespread propaganda and censorship against creation, intelligent design, and young-earth evidence, and
the endorsement of *only* evolution and billions of years, real and often
hidden *motives* should be considered. This calls to mind the international-
ly-famous evolutionist brothers from England, *Sir* Julian and Aldous Huxley,
and their candid responses in interviews and speeches some sixty years
ago (Aldous could've been a *Sir*, too; he declined knighthood). They were
among the leading voices for evolution and humanism in the world at that
time, around the centennial celebration of Darwin's book (1959). Sir Julian,
in his ever-stalwart commitment to explain things without God, said this
in the keynote address at that 1959 Darwinian Centennial:

---

[146] Ibid, Baskerville, p. 1

[147] Ibid, Sanford, *Genetic . . .* p. 5

> Darwin pointed out that no supernatural designer was needed;
> since natural selection could account for any known form of life;
> there was no need for a supernatural agency in its evolution .
> . . we can dismiss entirely all idea of a supernatural overriding
> mind being responsible for the evolutionary process.[148]

We now know Huxley and Darwin were wrong on this point—natural selection **cannot account for even a single form of life**—just tiny variations in *existing* forms of life, which is observed all the time (but never leads to entirely new body types or organisms, or *macro*evolution). Huxley called Darwin's evolutionary story "the most powerful and most comprehensive idea that has ever arisen on earth."[149] I'm not sure about that, but it is definitely the most powerful and influential idea in the last 200 years, as this book and real history demonstrate. On several occasions when commenting on his own and other like-minded humanists' opposition to objective consideration of creation arguments, with instead only the dogmatic and enthusiastic embrace of unproven evolution, Huxley admitted that Christianity, if true, *would interfere with their sexual mores and choices*. Talented writer, philosopher, and intellectual Aldous Huxley, Sir Julian's influential brother, was an early champion of psychedelic drug use (1950s), and had a no-hope worldview that shone forth in his writings. Aldous was a deep thinker; his negative worldview came from his realization of the ultimate ramifications of evolutionary materialism, akin to Friedrich Nietzsche's earlier *"God is Dead"* proclamation. Back to *motive*, though, he said:

> I had motives for not wanting the world to have a meaning
> . . . For myself, as, no doubt, for most of my contemporaries,
> the philosophy of meaningless was essentially an instrument
> of liberation. The liberation we desired was simultaneously
> liberation from a certain political and economic system and
> liberation from a certain system of morality. We objected to
> the morality because it *interfered with our sexual freedom.*
> There was one admirably simple method of confuting these
> people [Christians] and at the same time justifying ourselves

[148] Julian Huxley, *Issues in Evolution* Sol Tax, ed., Univ. of Chicago Press, 1960, p. 45

[149] Julian Huxley, *Essays of a Humanist,* Harper & Row, New York, 1964, p. 125

in our *political and erotic revolt*: we could deny that the world
had any meaning whatsoever.[150]

So, there you have it, from one of their most prominent 20th century mouthpieces. It is a sad and interesting study to reflect on the comments (and fruit) of some of the intelligentsia and artists of the early and
mid-twentieth century, besides the famous Huxley brothers, who were
caught up in the almost unchallenged evolutionary juggernaut. This
included most of the *Bohemian* crowd with their new-found relative
morality, centered in Paris after WW1. This included artists, philosophers,
social and political commentators, and writers from various countries—
including Ernest Hemingway—who described those days as "a Moveable
Feast." Hemingway was not totally ignorant of biblical statutes and precepts, as evidenced by occasional quotes and even titles from some of his
works (e.g., *The Sun Also Rises*). But he, too, seems to have almost reluctantly bought into the prevailing new materialistic worldview that swept
academia and the arts, but that ultimately led to despair and the dearth
of hope. In his later years, he admitted, "I live in a vacuum," and was only
61 when he committed suicide in 1961. Philosopher, social activist, mathematician, and evolution supporter Bertrand Russell, brilliant in only some
ways and rarely hesitant to criticize Christians, had many quotes on loneliness and despair with his similar no-hope worldview. Here is one I came
across years ago: "We stand on the shore of an ocean, crying to the night
and the emptiness. Sometimes a voice answers out of the darkness, but it
is the voice of one drowning, and in a moment the silence returns." That
same unfortunate life-sapping denial of purpose- or meaning- is undeniably more widespread today. Additional similar quotes could be offered
from other early 20th century intellectual/evolutionists relative to despair
and *motive* (especially sexual freedom); hopefully you get the point. Even
a century ago, it wasn't *just about* the supposed scientific evidence, not
even close.

Earlier it was pointed out that among the scientists, scholars, and pastors involved in the creation/evolution debate, it is pretty much accepted
that there are only two legitimate options, or explanations, of *how we
came to be*. If the strange, non-scientific idea that things and life forms
with incredibly complex programmed information somehow made themselves is discredited, and shown to be untenable, then realistically that
leaves but one option. That remaining option: there was and is a Designer,

---

[150] Aldous Huxley, *Ends and Means*, Chatto & Windus, London, 1938, pp. 269-270,
273

or *Creator God*. For most people, it logically follows that it-or *He*, from the Christian perspective-has the right to set the rules, or life guidelines, including on gender, sexual practices, and marriage. In such contemplation of this Creator-God option, it also forces (or at least nudges) every thinking person to consider the subject of future rewards and punishments.

Back to marriage, it bears repeating that in referring to Genesis 1: 27 and 2: 24, Jesus himself said, "*From the beginning of the creation* God made them male and female, [and] for this cause shall a man . . . cleave to his wife, and the two shall become one flesh."[151] Jesus, just 2,000 years ago, is stressing that the triune God made Adam and Eve right at the actual beginning (between six and seven thousand years ago), *not* 4.5 or more billion years ago *after* an imagined prior eons-old beginning, as today's evolutionists and even some compromising pastors teach and promote. Additionally, in answering those sluggish Pharisees with cobwebs in their brains (Matthew 19), He responds to their question concerning marriage and family-the most important of all human institutions-by quoting *Genesis*. He plainly regarded it as real history, and divinely inspired (He was there). The God of the Bible has many statutes, commands, and precepts which experience, history and even science clearly show lead to a better life and a safer, more productive society with more true liberty (not licentious-ness) and freedom. If one checks the various countries and nation-states throughout history, and their accompanying religion (moral compass), or the lack thereof, one can compare the cultural *fruit* on the tree. There indeed seems to be something to the notion that *culture is religion exter-nalized*. Take a look—with intellectual honesty—at those areas and nations of the world founded on the Judeo-Christian ethic, and especially since the Protestant Reformation, and do a quick comparison with others (see the following *Country/Religion/Cultural Fruit* comparison section).

This leads us finally back to morality, including sexual behavior and practices, and the possibility of divine laws governing such behavior. If his-torical-based Christianity—the only religion in the world (besides Judaism) that is actually *set in history* and *predicts the future*[152]—is really true, then the Bible's clear commands, precepts and restrictions on sexual relation-ships are vitally important. Current compromising liberal theologians and pastors notwithstanding, these are clearly laid out in Scripture, and viola-tions thereof are among the most egregious of sins (e.g., Genesis chapters

---

[151] The Bible, KJV, Mark 10: 6-8 (see also Matthew 19: 4-7)

[152] For a great introductory read on biblical fulfilled prophecy, consider *a Closer Look at Prophecy*, by Richard and Tina Kleiss, Search for the Truth Publications, Midland, MI, 2019.

6, 18, and 19; Leviticus 18; Romans 1:20-32, and I Cor. 6:9). Such was the predominant belief in historic Christendom and more recent Western civilization, and particularly the United States, until the last century and a half with the rise and domination of evolutionary materialism (or naturalism), in this postmodern age.

Homosexuality, lesbianism, transgenderism, gender dysphoria and the entire sexual genre is a touchy and awkward subject because it is personal. But as is becoming increasingly apparent with the sudden explosion of identity/gender confusion, legalization and promotion of homosexual marriage, transgenderism, and hormonal and surgical gender-change treatment, new challenges and problems abound. Amazingly, there are even demands by the political left, including some of the Democrat presidential candidates (2019-2020), for these procedures to be *taxpayer-funded*, offered, and performed on prison inmates and military personnel. Lonely, lost, and confused kids are seemingly everywhere; diminished national academic performance abounds; we have fast-rising suicide rates and destroyed families; less productive lives, a plethora of sexually transmitted diseases, and overwhelmed social services; and now a quickly growing dissatisfied number of young adults that regret their sex/gender-change decisions, and want to revert back, among the many new problems. Is this good fruit from solid, righteous policies? Have parents, advisors, psych docs and academics, as well as the mainstream media and political pundits on the left, seriously thought this through and considered *long-term* ramifications? Of course they haven't. Let's talk about it.

I'll not go further into the differences between men and women, which were mentioned earlier, and are well documented, other than to say that every human body cell (except enucleated blood cells) is either female (XX chromosomes) or male (with XY chromosomes). Thus tiny, new precious humans begin with fairly substantial sex chromosomal differences, with the X containing well over 800 genes and the Y having only between two to three hundred. After fertilization, a 1-2 day excursion down the Fallopian tube is followed by implantation of the miniscule human zygote in the hormone-prepared uterine wall, and this single unique cell begins an amazing process. Controlled by DNA, chemical communicators and hormones, and sometimes esoteric epigenetic factors, it divides by mitosis and begins the awe-inspiring, still incompletely understood (putting it mildly) replication, growth and differentiation process, to produce the various tissues, structures, and organs of the human body. Early in the process, certain genes are suppressed or turned off, varying in the two types, and males and females are on their way, with many built-in differences. The sex-determining segment on the Y chromosome, for example, is a gene that inhibits female development but induces male anatomical differentiation about six to seven weeks after fertilization. Maleness or femaleness is imprinted into *every one* of our 30 to 80 trillion (depends partially on body size) or so living cells, and no hormone chemical treatment, gender surgery, cross dressing or outlandish claims can change that.

There is no conclusive evidence that homosexuals are *born* or *made* that way from the get-go. It is interesting that there is now a growing number of homosexuals who are actually uncomfortable with that contention. The many individuals who have transitioned *to*, or *back to*, a heterosexual lifestyle and/or successful heterosexual marriage from a homosexual or lesbian lifestyle also give evidence against that claim. Homosexuality is a choice, and it is not natural. I saved a poignant commentary by a sharp young man, who was a Catholic seminarian, from a NARTH (National Association for Research & Therapy of Homosexuality) newsletter from 2004, which really touched my heart.[153] His unforgettable letter was addressed to a leftist psychologist, who was teaching one of his classes. Here is part of it:

> I cannot recall any particular moment or age in my life when I was willing to say, "I'm gay." Throughout my life I abhorred these feelings. Somehow I always knew that there was something wrong with homosexuality, but I could never put my finger on what it was. I guess you can classify my psychological state as ego-dystonic, because I wrestled (and still do today) with homosexual thoughts and feelings, yet only acted out by way of sexual fantasy and masturbation, never with another man . . . I see homosexuality as incomplete psycho-sexual development; as a matter of stunted emotional growth in the area which would have prepared me to enter the world of men. The idea is simple: at some point in my childhood years I defensively detached from my father and the masculinity he offered due to hurt in the relationship or separation, and instead bonded more with my mother, my primary care-giver. . . It is said it takes a mother to raise a child into a boy, but it takes a father to raise a boy into a man. And that didn't fully happen for me . . . For too long I have sought out this lost masculinity by homosexual attraction and desires, rather than through healthy friendships and relationships.

---

[153] See, http://www.narth.com/docs/change.html

He then describes the feeling of "my blood boiling," as he listened to the compromising seminary psychologist professor essentially defend homosexuality, political correctness, and relative morality, and he thus responded as follows:

> First, I feel betrayed by society in general, which is lost in relativism and subjectivism, and which tells me that I should accept myself "the way God made me." God made me a human person, in His image and likeness. God didn't make me gay. I can never accept that. Homosexuality is a result of the Fall. You do a great disservice to encourage someone to embrace a false identity. Second, [some Catholic] Church leaders have told me the same thing—that I should accept my "gayness," that I have "gay gifts" to offer the Church. This is more nonsense. A psychological disorder cannot offer anything but more problems . . . Homosexuals are not hard-wired to come out homosexual. That is a lie propagated by the gay agenda . . . The evidence for psychological root causes of homosexuality far outweighs the inconclusive biological research.

Remember, this sharp young man is directing this letter to the psychologist who spent significant time on homosexuality, while teaching one of his previous classes. You can almost feel the seminarian's wounded heart. Although well-written and diplomatic, his emotion and boldness gradually increase:

> You must figure out if you, too, have been "duped" by the gay lobby, robbing you of your objective judgement in this matter. What can you offer people like me? . . . There are many of us out here, that want to say that the gay agenda is a lie-it insists we accept something disordered that leads to immoral acts; it insists we were born gay; that we cannot hope to change. Now even you, yourself, tell us that it is impossible to convert the homosexual man to heterosexuality. You also bring us a false understanding of homosexuality and heterosexuality . . . [then, he boldly *teaches the teacher*] One is abnormal, the other normal; one is incomplete gender identity, the other complete; one is a lie; the other true . . . Homosexuality offers

me empty promises, to seek the masculinity of others by sexual means.

He mentions help and progress through Dr. Joseph Nicolosi and his NARTH organization, Church teaching through *Courage* (a Catholic spiritual support group), as well as through private confessors and close friends. He says he "has been healed greatly—maybe not to complete heterosexuality, but I'm getting there." Touching on fatherhood, he closes this heartfelt letter with a plea:

> Take a look at your son. Wouldn't you want what is best for him? What if he were gay? If homosexuality is treatable to a point where the person reaches a sense of wholeness about his life—that he is a man in every sense of the word—wouldn't you want that for him? By saying "there are no easy answers," you offer us empty words. There are answers in life; the problem is that people *just don't want to hear them.*

His closing sentence is so true in so many spheres of life. There *does* seem to be solid evidence indicating certain family and early peer group dynamics and relationships, or the lack thereof (family/social factors, if you will), which may predispose an individual to choices in the same-gender direction. In fact, dislike of parents and particularly of fathers is a common situation leading to the homosexual lifestyle. And now, some fifteen years after that poignant letter, the reality of successful change from this homosexual lifestyle to, or back to, heterosexuality and even heterosexual marriage is irrefutable, as mentioned earlier.

Former South Bend, Indiana mayor and Democrat presidential candidate Peter Buttigieg, who came out of the closet, so to speak, before his second mayoral term, recently unjustifiably slammed Vice-President Mike Pence, a strong Christian, while the mayor was commenting on homosexuality: "If me being gay was a choice, it was a choice that was made far, far above my pay grade. And that's the thing I wish the Mike Pence's of the world would understand. That if you got a problem with who I am, your problem is not with me—your quarrel, sir, is *with my Creator.*" Since Mayor Pete just publicly stated his belief in the *incompatibility* of Christianity and conservatism (Feb., 2020), and the Vice-President is irritatingly from Indiana, Mike Pence must be particularly bothersome to Buttigieg. His earlier comment is very problematic, though, as it denies scientific and social evidence, the role of personal choice, is borderline blasphemous, and ignores and denies Biblical truth or simply indicates *profound biblical ignorance on Buttigieg's part*. I mean- just for starters, Mayor Pete- have you read from one of the most important chapters in the entire Bible, from the most powerful book in the New Testament—Romans—in chapter 1: verses 20-32 ?! Or how about Genesis 19:4-9, or Jude 7?! No one can stand on God's Word if they don't know what it says.

Having a civil, evidence-based intellectual debate on moral issues (virtually impossible today-especially in an academic setting) seems to usually make those who "willingly are ignorant," or have reached a point where they are blind to the truth, more vitriolic and defensive in denial. Perhaps this is because such individuals choose to ignore reality by ostensibly believing something which they know, deep within their core, cannot be true.[154] After all, it is irrefutably part of our nature to lie and deceive. Authors Julie Von Vett and Bruce Malone describe it this way:

> The problem with lying to ourselves is that we actually know we are lying, so we must pile layer after layer of more absurd lies in a futile attempt to suppress the original lie. Homosexuality is a denial of the obvious. Human bodies are designed to fit together for reproductive purposes-male into female-like a key into a lock.[155]

It seems simply ridiculous, as well as indicative of our degree of cultural confusion, diminution of academic scholarship, and spiritual emptiness that this even needs to be pointed out. In my own experience with farm and hunting equipment, tools, and dental units and hoses, "male" and "female" adapters and complimentary parts are used and identified as such all the time. Where do the gender-confused and LGBT crowd think we got our model for those terms? Is it not stunning that we are at a point where some people now feel they can even *determine their own* sexuality, and in many schools kids are being taught that they can *choose* their sex, and that their sexual identity is the most important issue?! Is it not sad what many children are now exposed to, and the loss of the age of innocence? Even down to the kindergarten level, these youngsters are often taught that their body is irrelevant, and has very little or nothing to do

---

[154] J. Budziszewski, *What We Can't Not Know*, Spence Publishing, 2003, p. 158

[155] Ibid, Von Vett and Malone, *Inspired . . .* June 15th reading

with *who they really are*. Unsurprisingly, all of this can lead to confusion and serious mental and physical problems. This is not just my opinion; there is much supporting data of this. Abnormal childhood family dynamics and relationships as well as academic indoctrination are big parts of the problem, but the deeper foundational problem is the casting aside of our Judeo-Christian ethic and biblical precepts (moral absolutes) in favor of non-evidence-based evolutionary materialism. I hope you readers are starting to sense that as with so many moral or sexual topics, any discussion over gender-change questions, or homosexual marriage, can only be *sensibly* debated relative to the foundational issue of origins, or *where we came from*.

A tiny group of fallen, **not-able-to-be-fired**, *un*elected officials—the men and women of the **United States Supreme Court**—have stunningly already ruled that homosexual marriage is equivalent to biblical marriage between a man and a woman, as they struck down all bans on same-sex marriage in the United States in June, 2015. Backtracking a bit to 2003, the High Court then ruled that an existing Texas state law criminalizing sodomy, or "intimate sexual contact between two consenting adults of the same sex" was *un*constitutional. The Court surprisingly and illogically there reversed itself as it struck down its own decision- from just 17 years before- in the Bowers <u>vs</u> Hardwick case. In that particular case, the Supreme Court ruled *in support of* state laws when it declared that sodomy was *not* protected behavior and was *not* a constitutional right. **Why** the diametrically opposed **reversal** just 17 years later, in 2003? A legitimate explanation was never forthcoming.

The real answer: caving to the incessant pressure of the sexual revolutionaries and political correctness on top of the demise of biblical authority. The subsequent 2015 High Court ruling, as they once again overstepped their Constitutional bounds in and **engaged making/breaking law**, or unwarranted judicial activism while thwarting the will of the majority, could definitely be classified as a landmark decision. It was an obvious confirmation—with a large exclamation mark—of the complete and final loss and disrespect of biblical authority by the United States Supreme Court (due in large part to the evolution/no God worldview). And then, exactly two years later (June 26th, 2017), the same High court ordered all states to treat same sex couples equally to opposite sex couples in the issuance of birth certificates, thereby permitting legal adoption by same sex couples in all 50 states. Many exasperated Christians and heterosexuals wonder, *what's next*? Or, *is that all*?! The real answer: They're **just getting' started**. Same-sex marriage and parenting is *only the beginning*.

Even well before the infamous 2015 High Court ruling, the movers and shakers, so to speak, of the sexual revolutionaries put forth a significant statement. A manifesto entitled "Beyond Same-Sex Marriage," issued by prominent "lesbian, gay, bisexual, and transgender (LGBT), and allied activists, scholars, educators, writers, artists, lawyers, journalists, and community organizers," demands "legal recognition for a wide range of relationships, households, and families, and for the children in all of those households and families." Despite professions by "moderate" advocates that monogamous same-sex marriage is all they seek to legitimize, Stanley Kurtz argues, this document demonstrates that **the real agenda is *"to dissolve marriage*, not through formal abolition, but by gradually extending the hitherto unique notion of marriage to *every conceivable family type.*"**[156]

It is truly amazing when you think about it; not very many years ago, sodomy was **against the law** in all 50 states. Don't put too much faith in those *unelected* judges. For example, it was also the black-robed dudes on the United States Supreme Court that **absolutely stripped** the power, enforcement, and intent **from the three constitutional amendments** that were passed *after* the War Between the States **to bring former slaves and African- Americans equality**, with the unfortunate and terrible reality (over a century ago) that those amendments were made virtually useless. With higher academia's new enthusiastic evolution-teaching of *favored* and **inferior** *human* races during that same time period (early 1870s-late 1930s), **racism** sadly got a big shot in the arm. Never forget, those High Court justices are fallen human beings, with "feet of clay," just like the rest of us.

So what else? Recall my earlier contention that the downfall of every great nation-state is preceded by a *change in the language*. That's been happening so fast in our country in recent years it makes your head spin. As Stephen Baskerville points out, pundits, professors, and even many intimidated pastors have substituted legal jargon and political correct words for Christian terminology, as they have promoted or allowed political ideology to replace Christian sexual morality.[157] He points out that we hear terms like sexual harassment, sexism, and misogyny quite often these days. History shows that sin always advances without a restraint by some sort of salty righteousness. Is it just a matter of time—as just hinted at in the "manifesto"—before it is legal for a man to marry ten women, or three men? Or how about a woman marrying ten other women, or perhaps even

---

[156] Ibid, Bask, p.97; and citing Stanley Kurtz, "The Confession," National Review Online, 31 October 2006, http://www.national review.com/articles/219092/confession/Stanley-kurtz

[157] Ibid, Bask, Intro & chapter 1

her own son? Should there be any restrictions to man-boy sex, which was not uncommon in ancient Greece and Rome? How about even human-animal sex? In a secular society with no God or moral compass, who is to say what's right or wrong? Who sets the limits? Are the people simply subject to *arbitrary* rulings by those in power? Keep in mind those in power—i.e., the government—is irrefutably the dominant partner and the intimidating, hammer-like enforcing arm of the radical feminists and other sexual revolutionaries. Central to these questions, as well as stemming further descent into the cultural abyss, is coming up with the right answer to the already-emphasized question, *"Where did we humans come from?"* Consider this perspective:

> Since our entire educational, media, and museum systems promote the idea that we came from single-sex organisms, there is no basis in reality to justify elevating the marriage between a man and a woman above any other arrangement. Examples can be found in the animal kingdom to justify any possible sexual arrangement, and we have trained children for decades that we are just another type of animal. BUT, if man and woman really were created for the purpose of completing each other as stated in the Bible, (obvious by the "key-in-lock" structures of our physical bodies and complimentary talents/abilities), then **any other arrangement in marriage is simply a distortion of truth**. The whole gay marriage debate simply disappears. We need to get back to fighting this cultural "trainwreck in progress" from an origins and biblical basis. When the evidence supporting Genesis is acknowledged, the justification for relative sexual morality simply disappears.[158]

Indeed, this volatile issue is another example illustrating the contentious *origins/long ages* (or *deep time*) debate **is not an unimportant side issue**. The important recommended reading passage for confused South Bend Mayor Pete, in Romans 1: 19-28, clearly portrays what happens when a people- group casts aside the Creator-God and worships the *creature*, or creation (same word in the Greek), more than the *Creator*. We see this all around us in our post-modern world. Bible scholar and scientist Dr. Henry Morris commented on what *always* follows, this from 25 years ago:

---

[158] Ibid, Von Vett and Malone, *Have You Considered* ? Dec. 1ˢᵗ reading

The descent into evolutionary paganism is always soon followed by gross immorality, specifically including sexual perversion, such as described in Romans 1:26-29. Ancient Sodom was so notorious for homosexuality that its practice has long been known as sodomy (Genesis 13:13; 19:4-9). The practice became so widespread in ancient Greece that it was considered normal and even desirable. Other examples are abundant and, of course, it is quickly becoming accepted—even encouraged—here in America. Not surprisingly, this was preceded by widespread *return to evolutionism* in science and education.[159]

Back to new cultural challenges and complaints related to the modern day gender fluidity/transgender issue and promotion, we now have a growing and awkward problem with transgender athletes. Have we lost our collective ever-lovin' minds? It is **not fair**, like the case of the transgender high school wrestler who twice won the girls Texas State Championship, who had regularly been taking some serious testosterone injections in transitioning from female to male. Do you think the testosterone maybe, possibly, made her stronger? Or how about the men, who tend to be unar-guably bigger and stronger physically than women, claiming to be (or *identifying as*) women, winning world-class running and cycling events? Do you think the medically-confirmed larger amount of muscle mass, greater lung capacity, or larger number of red blood cells (think: more oxygen/energy/power), or perhaps the bigger hearts found in men might have anything to do with it? Even the Boston Athletic Association now allows men who just *identify* as women to register and compete as women in their pop-ular American footrace. The claim by the left and the transgender crowd that gender choice is optional, or subjective, and personal—as well as something entirely different from one's biological sex, is incomprehensible. Biologically, there are not only anatomical and morphological differences, but also **physiological differences** between men and women, that are in large part caused by sex hormone differences, which in turn are caused by sex *chromosome* (XX or XY) differences. Complaints and frustrations with gender cheating in athletic competitions are increasing, and it's **not fair**. The incessant calls for *tolerance,* along with omnipresent political correct-ness run amok is tiring, is it not? (*Hint: that's part of the left's strategy, *to wear. . . you. . . out.*) It's getting senseless out there; we're at the point

---

[159] Dr. Henry M. Morris, annotations author, *The Defender's Study Bible,* World Bible Publishers, Inc., 1995, Romans 1:26 annotation, p. 1231

now (end of 2019) where claiming there are only two genders or sexes is considered a *hate crime*! More people need to stand strong for common sense, good health, and fairness. Don't always go along to just get along.

How about sensible counselors, psychiatrists, and psychologists (yes, there are a few good ones) that try to do that very thing on some of these youth gender confusion cases? Well, they'd better be careful; they may be risking their job or career. An example would be what happened to University of Louisville professor and psychiatrist Allan Josephson, who chaired the school's division of child and adolescent psychiatry for over fourteen years, and had excellent performance reviews, before the following incident. It all started a couple years ago (late 2017) when Professor Josephson spoke on a panel discussion about gender dysphoria and children, where he challenged the idea that gender *identity* should take precedence over the child's *real* biological gender, or their sex at birth (XX or XY), as such an endorsement, or promotion, was "counter to medical science." Gender *non*-confused, solid, sensible, people across the fruited plain **do not need to be told** that it's *counter to* medical science; *of course it is!* But then, this psychiatrist (who obviously really cares about children) apparently just went too far, for the hovering academia/media thought police and transgender crowd. The professor wisely suggested that parents, after a loving sit-down and patient listen to their children, essentially get on the same wavelength and use their common sense and wisdom in guiding their child to align with his or her biological sex. Again, most American's probable response to that advice: "*Well, absolutely; very good counsel, Doc.*" However, a few liberal colleagues complained, Josephson was demoted, and then *un*justifiably fired in early 2019. He's not the only case out there, either. Attorneys for Alliance Defending Freedom have filed a lawsuit against the University of Louisville on the professor's behalf, accusing them of free speech violations.

On *the left* it is now free speech *only if it agrees with their worldview*. Are you readers not concerned that professional counselors and doctors, who question or buck the leftist party line on opposite-gender and identity claims and change (especially on the young), because of solid health/medical reasons, common sense, and occasionally moral reasons, are quickly and aggressively targeted and attacked? In the minds of many of us in flyover country *and* the medical establishment, it is not only ridiculous, but wrong and borderline criminal for parents, certain physicians, and counselors to promote and encourage these radical, life-changing, usually permanent changes to their young bodies, instead of going slow and carefully considering social, nutritional, psychological and mental health deficiencies and issues. Do not be intimidated, folks.

Millions of hard-working Americans and Christians are sick and tired of the *mis*characterization and shaming—by the secular-left minority—of themselves and all others who **don't** condone or support gender fluidity, homosexuality, lesbianism, and/or transgenderism, not to mention socialism, man-made climate change, and more. This widespread group of frustrated people (myself included) represents all of our ethnic groups, many with a faith-based worldview, but some without. They are presently still in the majority in this nation, and yet are classified as haters, homophobes, ignorant (of science, history, etc.), bigots, racists, and worse. What really irks this group is the liberal left's refusal to state the people's **real objection**, which is *moral corruption, in opposition to Judeo-Christian principles*, and the sexual revolutionaries' *attempted destruction of the family*. Do not doubt that last claim; even in the 1980s, well-known feminist academic Judith Stacey (and many others), a supposed expert on gender and queer studies, promoted "good riddance to the traditional family."[160] That goal by the left in the last few decades has been pretty successful. Homosexual activists have successfully followed the radical feminists' lead. They've now attained legality, equal and civil rights; emboldened, they want more. "Homosexuality today is more than a private personal preference; for militant activists, it is a political ideology that aims to use sex to transform society in the homosexual image."[161] Turning again to author-political scientist Stephen Baskerville, he draws from Michael Brown on their *real* long-term goal:

> "Being queer is more than setting up house, sleeping with a person of the same gender, and seeking state approval for doing so," says the former legal director of the Lambda Legal Defense and Education Fund. "Being queer means pushing the parameters of sex, sexuality, and family, and in the process **transforming the very fabric of society**."[162]

---

[160] Ibid, Bask,, p. 37; from Judith Stacey, "The New Conservative Feminism," *Feminist Studies* 9, 1983, p. 570, quoted in Bryce Christensen, "The End of Gender Sanity in American Public Life," *Modern Age*, vol. 49, no. 4, Fall, 2007, p. 409

[161] Ibid, p. 37

[162] Ibid, p. 37, taken from Michael L. Brown, *A Queer Thing Happened to America* (Concord, NC: Equal-Time Books, 2011, p. 37

At this tipping point in American history, as the cultural chasm deepens from coast to coast, Americans of faith and family values need to understand the seriousness of the battle, and the sexual revolutionaries' real agenda and long-term goals. The central plank was revealed decades ago in the 1978 Gay Liberation Front *Manifesto*:

> "We must aim at the **abolition of the family**, so that the sexist, male supremacist system can no longer be nurtured there."[163] Despite disclaimers similar to the feminists', there can be no doubt that the homosexual agenda remains fundamentally hostile to the family, to heterosexuality, in general, and **above all to traditional masculinity**. "The oppression of gay people starts in the most basic unit of society, the family, consisting of the man in charge, a slave as his wife, and their children on whom they force themselves as the ideal models. The very form of the family works against homosexuality." These people are not merely asking to be left alone. Hostility toward parents and parental authority, **especially fathers**, is palpable in homosexual literature, and **the aspiration to eradicate them is undisguised**. This hostility extends beyond what any particular parents *did* to encompass what parents in general are: married, heterosexual couples who acknowledge differences between men and women . . . Feminism and homosexualism thus target not simply men and masculinity but fathers above all, the embodiments of the hated "patriarchy" and the "system" by which "men dominate, oppress, and exploit women."[164]

Unfortunately for people of historical Christian faith and biblical family values, the sexual revolutionaries have made great progress in the last forty years in pursuit of this goal. The abuse and suppression of men and fathers and the destruction of the American family is well underway, initially led by the radical feminists and academic elites, with quick support from leftist Democrat politicians, Hollywood, and the secular mainstream

---

[163] Ibid, pp. 37-38, taken from Fordham University internet site: http//www.fordham.edu/halsall/pwh/glf-lon-don-html.

[164] Ibid, p. 39; taken from Sylvia Walby, quoted in Anne Barbeau Gardiner, "Feminist Literary Criticism: From Anti-Patriarchy to Decadence," *Modern Age*, vol. 49, no. 4, (Fall 2007), p. 93

media. Are you starting to get at least an inkling of understanding of where, from whom, and *why* the term **"toxic masculinity"** originated? It is a fabricated term by the radical left to facilitate their political agenda, power, and secular worldview. This comprehensive attack on men and historic American manhood would not even be possible without the sad replacement of biblical authority with ungrounded evolutionary materialism. The ongoing successful effort by the radical feminists and homosexualists at creating more legislation and tools, and to increasingly enlist the strong-arm power of state and federal governments to criminalize male sexuality and men, is relentless and serious indeed. For more detailed and stunning information on this topic, consider reading Part 3 -*Criminalizing Sex: The New Gender Crimes*- from the culture-war classic *The New Politics of Sex* by Stephen Baskerville.[165]

To frustratingly reiterate, these controversial behaviors, practices, and policies that many American citizens **do not support** were, until recently, considered **immoral, unthinkable, some were actually non-existent, *and illegal***. Without an activist federal judiciary, particularly the **United States Supreme Court** greatly over-stepping its constitutional bounds, we would be nowhere near the cultural mega-mess we're in today. The High Court and American judiciary, just like the entire public and higher education system, are now based on the secular, evolution-based materialistic worldview with its required *changing*, or *evolving* laws (no absolutes) and culture-disrupting relative morality. We've got a problem.

It should be mentioned that the word *homophobia* was defined years ago by the *Kinsey Institute New Report on Sex* as the "fear, dislike, or hatred of homosexuals."[166] Well, I don't agree with their lifestyle choice, but I've never experienced fear, hatred, or dislike of homosexuals. In all honesty, I admittedly have seen some on television news-clips and parades, usually screaming, yelling, or exposing their private parts that I would probably not favor. I've personally known more than a few, have definitely had some homosexual and lesbian health and dental patients, and in years past even performed dental work on several HIV-positive individuals and helped them as much as I could. In *my* personal experience, I got along just fine with *all* of them. I also had several devoted lesbian dental patients, two of whom somehow tracked me down after I moved out in the country twenty years ago after selling my Charlotte, N.C. dental practice. I purchased a

[165] Stephen Baskerville, *The New Politics of Sex-The Sexual Revolution, Civil Liberties, and the Growth of Governmental Power*, Angelico Press, Kettering, OH, 2017

[166] June M. Reinisch, dir., *The Kinsey Institute New Report on Sex,* St. Martin's Press, New York, 1990, p. 147

smaller dental practice from a retiring gentleman some 25 miles away, and said nothing about it to any of my Charlotte patients (part of the deal; I didn't want to steal any of my old patients from the young buyer, whose success I desired). But somehow, these two gals eventually tracked me down after a year or two, and insisted that I not only accept them back as patients—they didn't mind the now 50-minute drive—but *scolded* me for leaving them (even though I sent all of the patients an appreciative thank you letter, while edifying the new dentist). Our colorful relationship, with much good-spirited joking, thus continued.

Back to the "**phobia**" of homophobia; it actually means *an irrational fear.* The only thing or person I've ever had an *irrational* fear of was running into a black mamba while bow-hunting in southern Africa (and that may have been rational). A big problem in our current cultural divide is that the gay/transgender rights movement, with their academic commandants and water carriers in the mainstream media, considers anyone who opposes *anything* they do or stand for, or denies the *complete acceptance* of their homosexual, lesbian, or transgender lifestyle as *healthy and normal*, as nothing more than a racist homophobe. This is preposterous, and inhibits serious and honest discussion of the nuts and bolts, facts and negatives, long-term ramifications and so much more relative to these issues. If one understands their just emphasized true long-term agenda, the virtual impossibility of civil, serious discussion becomes much more understand-able. Such a pow-wow would include consideration of research findings indicating a suicide rate 8 to 10 times higher than the rest of the population among transgendered people, as well as health parameters and concerns and longevity of homosexuals, and so much more.

It would also include an honest discussion of this troubling new phe-nomenon that some are calling *rapid-onset gender dysphoria*. Because of promotion by LGBT activists, internet perverts and sexual revolutionaries, along with the usual spiritual void in their lives, many young teens—espe-cially girls—are suddenly declaring they are lodged in *the wrong body*. Interestingly, such stunning claims often occur in small groups, or *clusters* of teens. This hints at the power of peer pressure, and leads to the subse-quent quick request for cross-sex hormones, which could lead to a life-long dependency with worrisome side-effects as well as possible irreversible surgery, and a cascade of mental/emotional problems. Early research on this new movement interestingly indicates many of these youth come from *non*-Christian/*non*-conservative families, with left-leaning views on sexual practices. Significantly, in an academic study by Brown University's

Lisa Littman (fall, 2018)[167], from her *parent reports* and interviews, she found that **more than 60 percent** of parents of the discontented gender-dysphoric young people admitted that their child had been diagnosed with some type of **mental disorder prior to** their gender dysphoria revelation. Across the pond, it is reported that **70 percent** of those with gender dysphoria suffer from some form of mental illness during their lifetime, according to the *British Journal of Psychiatry*. Thus the actual picture so far, relative to this new rapid-onset disorder, is *not* one of solid, confident and grounded, productive, upbeat, faith-based teenagers suddenly—out-of-nowhere— making a careful but potentially life-changing decision. These troubled adolescents—on the contrary— that are dissatisfied in their own skin suffer from anxiety, depression, in some cases possibly nutritional deficiencies, and perhaps a spiritual emptiness as they search for meaning (and love) in the wrong places. Recall my earlier contention that we are more accurately described as *spiritual beings with a human problem*, rather than human beings with a spiritual problem. Reach out and help, or mentor these young folks, when you can . . . and give yourself away. . . but do be careful; you might get thrown in jail.

Understand that *despite* academic, media, political, and corporate endorsement of the LGBT agenda and lifestyle (often just for money, sales, and votes—equivalent to *power*), heterosexual, conservative, and/ or people with a faith-based worldview have the right to challenge their stifling group-think, and especially their **demands to approve** their behavior. Don't be fooled or intimidated by the incessant pressure to conform, or the demonstrations and activism, or the sometimes unwatchable Gay Pride events and parades, or the supposed desire for more *rights*. As already alluded to, they really don't want or need any additional laws or legislation; they *want you to like them and approve of their lifestyle*. A Catholic cleric accurately described the situation earlier this year, when I was visiting Ireland. I jotted it down; this may not be word for word, but its close: "In the 21st century it's not homosexuals who are persecuted, it's those who think their lifestyle wrong and say so. If you do you're probably guilty of "hate crime" and liable to be disciplined by your employer or denounced if you're in public office." All people, from the Christian perspective, are made in God's image and deserving of love, respect, and help, but people need to stand their ground for common sense, good health, and *right*. It's sometimes a matter of life and death for these young people—some of them are dying inside—physically and/or spiritually.

---

[167] Lisa Littman, M.D., MPH, Parent reports of adolescents and young adults perceived to show rapid onset of gender dysphoria, PLOS I ONE 13(8): e0202330. doi:10.1371/journal.pone.0202330

Back to the new issues—gender confusion, gender fluidity, 'non-binary trans' . . . and whatever—what would have been puzzling, head-scratching, unbelievable, "you've got to be kidding me!" concepts just a few decades ago. I mean, just coming at you from *Realville*—here in the Carolina pied-mont—considering most human beings, *what could be more obvious than gender*?! Why is it suddenly such an issue and why is there a rejection of gender? When you encounter and deal with the average human being in day to day work and life, what is initially more obvious than gender? When you notice body shape and structure, or clothing and shoes, and usually the hair- it is usually obvious. What about the pitch and nature of their voice, and/or the *way* they talk? How about their mannerisms and the way they walk? How about their eating habits? It is usually *apparent* what their gender is. But as discussed briefly in the early part of this book, the world out there—at least the left-leaning part of it—denies that those distinc-tions *even exist*. How can this be? How can they reject obvious biological gender? We don't need the politically-correct academic pseudo-esoter-ical explanation. The simple answer, that was alluded to earlier: they've *already rejected something far more important and obvious: that God exists*. Again, Romans 1:20 says all people are without excuse for their *non*belief in the Creator-God, because the evidence for Him is all around them (they "willingly are ignorant"). One could legitimately make a very serious case that the two most obvious things on this entire planet are 1) that God exists, and 2) that man is tainted; there's something wrong.

Last year Dr. Frank Wright, President and CEO of D. James Kennedy Ministries, described our gender/sexual dysphoria and confusion in our frustrating cultural impasse. He said:

> This is an issue that defies reason and common sense, with individuals demanding special privileges based on a seemingly arbitrary and preferential gender identity—as opposed to the natural and scientific understanding of gender as a straightforward function of biology. This is the ultimate "me-centered" expression of a world without truth:

> *I demand this freedom to be a man today, a woman next week, and a third thing the week after. And I demand this because, well, because I want it.*

This issue goes far beyond mere personal preference, because these claims of special privilege involve the safety and welfare of our children in their restrooms and locker rooms. Yet, here also, the response is the ultimate narcissistic expression:

I don't care about your children. This is what I want. This is what I demand.[168]

This anti-Christian, relative morality worldview of the progressive left— what Francis Schaeffer referred to as only *material energy*— has indeed produced some strange, regrettable, distasteful and even horrific cultural fruit today and for the last century.

Ending on a lighter note, though, occasionally we discover some almost *comical* residue here and there. How about the recent example of a teacher who, even though she switched to "being a man," was very agitated and *ill* (among southerners, means *seriously irritated,* or downright *mad*) because she ("he"?) still had some lower hemorrhagic effusion—a menstrual period—and the gal (or, guy?) strongly insinuated that . . . well, it just flat-out wasn't fair! Consequently, we now have in this confused culture human beings who have a baffling, unwanted menstrual period— when they don't identify as female—a frustrating social, biological and cultural frontier, indeed, in this challenging quest for menstrual neutralization or equity, with no mentors or guides down this heretofore untraveled path. Perhaps this particular individual is not really man *or* woman—possibly a new hybrid—more properly considered a "genderless humanoid dichotomy," or perhaps a "non-binary trans-teacher." Are you confused yet? All of this stuff creates unforeseen ancillary needs and expenditures, too, similar to when the U. S. Navy spent big money to re-outfit their fleets to accommodate female naval personnel. Multiple sources have reported at some colleges and universities the installation of cute tampon disposal containers (usually light coral, pink, or mauve) and other gender helpful/ neutral signs and apparatus in the men's bathrooms. I've heard nothing thus far on condom machine availability, color change, or policy changes, but have heard rumors of other soothing amenities in the works.

The truth can be stranger than fiction. Just a short time ago, you simply couldn't have made this stuff up.

---

[168] Frank Wright, Ph.D., *A Deadly Confusion*, from *Impact* Newsletter, D. James Kennedy Ministries, Sept. 2018, p. 1

32

GOODBYE, MR. BLACKSTONE

WILLIAM BLACKSTONE WAS a brilliant eighteenth century English jurist, judge, and law professor who is most famous for his classic four-volume treatise *Commentaries on the Laws of England*, often simply referred to as *Blackstone's Commentaries*. Blackstone's magnum opus was introduced in 1766, and it quickly became *the* law book in Great Britain, the American colonies, and then the U. S. Senate and legal profession of the new United States of America. His *Commentaries* and legal principles were based on biblical precepts and the almighty Creator, and Blackstone identified two streams of natural law: the "Laws of Nature," and the revealed law of the Scriptures, which were borrowed and embodied in our own Declaration of Independence: "the Laws of Nature and Nature's God." Part and parcel of this was the belief in *unchanging* divine laws and moral absolutes (real truth). I can even remember, as an inquisitive undergraduate student, wondering what this Blackstone guy must have done, to get his name etched in stone on law school and court buildings, in various locations, when I was a lost undergraduate student. As an awestruck teenager on my first trip to Morgantown, West Virginia, for a college football game, I noticed the name BLACKSTONE impressively engraved in stone, for all to see, up high across the front of the School of Law building as we drove past it. Of course back then I had never heard of him, but figured he must have been a pretty significant dude. I can confidently now state that it would be difficult to overestimate the importance of *Blackstone's Commentaries* on America's first 150 years. In jurisprudence, legal training, and law school philosophy and instruction, William Blackstone was irrefutably *the man*.

This all started to change at least by the early 1900s with the rise of Darwinian evolution, and the gradual replacement of the Founder's (and Blackstone's) biblical natural law philosophy, as relativism, or legal *positivism* spread forth. Thus we were faced with the newfangled idea that law is not, or *should* not be based on bedrock *unchanging* laws, truths or moral absolutes . . . because . . . well . . . they're simply aren't any. After all, if we're just evolved animals, the Bible's history is disproven, therefore its principles and precepts are mostly irrelevant as well, and thus law has to change with the times—or *evolve*— as well. As lightly mentioned earlier, Harvard Law School Dean Christopher Columbus Langdell championed Darwin's new evolutionary worldview, and introduced relativism applied to law— *positivism*—in the 1870s, as he and others quickly and enthusiastically applied Charles Darwin's new evolutionary *change* theme to law school teaching and jurisprudence. Langdell unfortunately introduced and promoted the *case-law* study practice, whereby **students would focus on judges' decisions rather than the Constitution**. What a massive problem—a plethora of *bad fruit*—we have today because of this signal change in legal philosophy. These ideas were particularly embraced and promoted by another Harvard guy, atheist Oliver Wendell Holmes, Jr. (1841-1935), in the first third of the 20th century. Holmes served on the Supreme Court for thirty years (not retiring until age 90), was very influential and admired by progressives, but was possibly **the most anti-Christian jurist of his era** as he was not only an evolutionist but espoused a form of moral skepticism, and opposed the biblical and Blackstone-endorsed doctrine of natural law. Perhaps you can get a better peek into Holmes' heart from his own stunning words: "I see no reason for attributing to man a significant difference . . . in kind . . . from that which belongs to a grain of sand." Wow, and that dude was on the Supreme Court for thirty years?!

Have you been frustrated in recent decades by *elitist* and *activist judges,* as many other Americans have? These are *unelected* judges who thwart the will of the people on a regular basis, ignore the Constitution, and instead of objectively interpreting the law, they take on the *legislative role themselves*—and **make law**—from the bench. **Thus they disobey the law themselves**. How can this be tolerated? Government by judges is unconstitutional and un-American. Although our country is in the midst of an obvious culture war, liberals—especially the far left kind—still struggle to get elected. Consequently they push their worldview and secular agenda through the court system. This is an important example why the origins debate—*creation verses evolution*—is not simply a side issue. American ground-floor evolution proselyte and Harvard Law School Dean Christopher Columbus Langdell—starting in the 1870s— encouraged and endorsed by liberal Unitarian Harvard President Charles Eliot, boldly taught and embraced the following idea: laws must evolve since man evolved, and **judges should guide both this important evolution of law *and* the Constitution**. This country-and-culture-changing worldview, as real science along with this book and many others illustrate, is based on the false foundational assumption (or *primary axiom*) of Darwinian evolution. Much cultural damage has been done, with this shift away from *a divine imperative for ethics*, to relative morality and unjustified and unconstitutional influence and power by secular academic and judicial elites. In recent years, the Law School powers that be finally and formally announced that they are **not** going to use the heretofore considered classic *Blackstone's Commentaries* anymore. The classic *Federalist Papers* have fallen out of favor as well. This is sad indeed, and basically a statement that (in their exalted opinion) there is no absolute truth, everything is simply relative, and standards and laws must *evolve* with the times.

Additionally—especially for you hayseeds, church-goers, and Walmart shoppers in flyover country (like me)—we're given the message that the academic elites and the black-robed judges on the high court know best what is best for us. Incidentally, over a year ago (Oct., 2018), the true character, backbone (or the lack thereof), and leftist political leanings of America's Law Schools were flagrantly revealed in a hastily constructed joint letter to the U. S. Senate, by over 2,400 of their law professors. This letter slammed highly respected Supreme Court nominee Brett Kavanaugh as having a "lack of judicial temperament." This was based on his emotional defensive reaction, at his second Senate pseudo-confirmation hearing, to the totally unjustified attempt to destroy his character, career and family by the Democrat Party. For many citizens, irrespective of political leanings, that televised hearing was one of the most uncomfortable spectacles ever experienced on American television.

Judge Kavanaugh's powerful and emotional response to the fabricated charges and obvious set-up actually saved the day for him, and *was to be expected from an innocent man with intestinal fortitude*. This *loss-of-face* letter from the U. S. law professors came forth **just a month after** the American Bar Association gave Mr. Kavanaugh the **highest possible rating** for professional competence, integrity, and judicial temperament. The legal profession is *not* neutral politically; in recent decades in our presidential elections, they have overwhelmingly financially supported the Democrat (the more secular and liberal) candidate. Our Founders, early statesmen, and other great lawyers, like the Christian patriot, formidable orator, and outstanding barrister Daniel Webster-men of character who helped change the world-must be rolling over in their graves. We recently struggled out of the twentieth century—the bloodiest in world history—with a lot of *bad fruit* from the trees of atheism and evolutionary materialism. This is a huge part of how we got in the mess we're in today, with the effect on the legal profession and the rise of unconstitutional judicial activism. Have you opened your eyes and ears and checked the fruit lately, in our confused culture? The *origins* debate is not just an unimportant side issue; it relates to everything.

Oh, and how about West Virginia University, and the formerly greatly-honored BLACKSTONE in stone, on the old Law School building? Unfortunately, the great Christian legal scholar is just an ignored nobody now; I've recently been informed his name there has apparently been smoothly covered over with mortar and obliterated. It would not surprise me. That university, like most others, is now undergirded by secular evolutionary materialism and relative morality. It was just named by *BestColleges*, in partnership with *Campus Pride*, as one of the "Best Colleges for LGBTQ Students" for inclusive programming and outreach efforts in an educational environment, and for the third year in a row. According to the fall, 2019 *WVU Magazine*, Chris Mayo, the director of their LGBTQ+ Center, and a professor of women's and gender studies, was named a recipient of the regional 2019 LGBTQ Leadership Award from the National Diversity Council. Additionally, the university was recognized with a 2018 Higher Education Excellence in Diversity (HEED) Award from *INSIGHT Into Diversity* magazine, for the fourth year in a row. My, my . . . how things have changed in Morgantown since I was there, and the things that are now celebrated and honored. It seems in certain locales (especially those with a lot of brick and ivy) truth is *no longer* timeless and unchanging . . . it evolves. I'm really missing Mr. Blackstone.

33

MOSES, JULIUS WELLHAUSEN, AND DISPROVEN GERMAN *HIGHER CRITICISM*

IN THE MID-EIGHTEENTH century, French physician and writer Jean Astruc was one of the first persons of note to challenge the long-accepted authorship by Moses of the Pentateuch and particularly the book of Genesis. Astruc tried to apply techniques of textual analysis to the book of Genesis, and came up with the idea that it was developed or written from several sources or manuscript "traditions," an idea I and probably many of you were confusingly taught in college, and is referred to as the *documentary hypothesis*. The book Astruc described this idea in was published anonymously in 1753. Astruc promoted the idea that two different writers wrote the two creation statements in Genesis 1 and 2, a heretofore unheard of claim the previous two millennia, and based his idea on the different names for God used in these first two foundational chapters.

This idea was rebirthed with a vengeance a century later by Darwinian evolution-influenced German scholar Julius Wellhausen, who developed it into the complicated "documentary hypothesis" on the authorship and origin of the all-important book of Genesis. Wellhausen reasoned— in his evolutionary, long-ages/primitive man worldview—that Moses and contemporaries of circa 1,300 B. C. didn't even know how to write. Consequently the Pentateuch (the first five books of the Hebrew Bible (Torah) or Old Testament), had to have been written much later, by four (or more) different sources or groups, usually referred to as the J, E, D, and P sources (supposedly representing the Jehovist, Elohist, Deuteronomist and Priestly documents).

I myself have forgotten a lot of the teachers, classes, and subject matter of many of my college undergraduate classes (give me a break, I'm in my sixties now), but I will never forget the Old Testament Studies class I took as a twenty year-old, where we studied Wellhausen's "German Higher Criticism" and how confusing and, at least to me, illogical it was. I was not even a Christian then (although I assumed I was), and rarely went to church—except at Easter and Christmas—but after taking that class I remember clearly thinking . . . how can anyone believe this book (the Bible), if this J, E, D, & P stuff is true because if it is, then the Bible cannot be trusted as written. I remember being surprised at having kind of a sinking feeling in my gut, even though I was barely a "lukewarm" (see Rev. 3:16) pseudo-Christian at best. A lost and confused young man back then was I, and even more so after that particular class.

Wellhausen and other like-minded European intellectuals' assumption that Moses and the other folks around 1250-1400 B. C. could not write is preposterous; they falsely assumed that Moses' era was still a somewhat primitive and non-advanced time period in man's cultural "evolution." The plethora of architectural, engineering and archaeological findings, in addition to records and tablets, from the Near and Middle East clearly shows not only phenomenal mathematics and engineering knowledge and skills, but writing and language skills in various nation-states, people groups, and kingdoms-some a millenium *before* Moses time.[169] Archaeologists have even discovered a library in the location of the ancient city of Ur, which was Abraham's home before his trek to Canaan. Discovered at Ur were thousands of stone *books* on not only scholarly subjects but ordinary mundane material, indicating that *the 'average Joe'* could even read and write. This was a good 4,000 years ago and was some *1,000 years before Moses.*

The use of *toledoth*, in the eleven delineating colophons, or "signatures," so to speak, found in Genesis relative to authorship ("*these are the generations of* . . . ") is not even considered, or simply ignored by Wellhausen and other critics (see more on *toledoth* in Appendix C). Consequently the historical and theological damage was done in the late 1800s, as Wellhausen and other German "rationalist" theologians, catalyzed by Enlightenment ideas and Darwin's new evolutionary worldview, nearly destroyed the Christian church in western civilization. Some notable Jewish seminary professors, leaders, and rabbis were strongly affected as well, including influential Lithuanian-American Mordecai Kaplan (profiled recently in an excellent commentary by Marvin Olasky).[170] The message thus conveyed: the Bible could not be trusted, but human

---

[169] For some great information on the brilliance and intellectual capabilities of ancient man, consider these two stunning and beautiful books: *The Genius of Ancient Man*, general editor Don Landis, Jackson Hole Bible College, 2012; and *Brilliant*, by Bruce Malone, Search For the Truth Ministries, 2014

[170] *Vanishing Dance*-Lessons on the Decline of Liberal Judaism, by Marvin Olasky, WORLD Magazine, September 28, 2019, p. 64

reason—particularly as exercised by theological and academic elites—could be. According to the German rationalists, led by Wellhausen, the supernatural creation of the universe, and Adam and Eve in the beginning, followed by all of the other early subsequent signal events of Genesis were just stories, allegories, or myths, that perhaps can maybe teach lessons or somehow illustrate a truth principle, but are not to be taken literally (even though Genesis is *clearly* written as an historical narrative).

This bludgeoning of the all-important Genesis text *fluidizes* the scriptures; one can then interpret them any way one desires. This calls to mind Jesus' well-known one-on-one conversation with Nicodemus (John, Ch 3), who was a member of the Sanhedrin and perhaps the greatest teacher in all of Israel, where God tells him "If you don't believe the earthly things (like creation, the cataclysmic Flood, the Confusion of Tongues at Babel, etc., which can be mostly proved-RJ), you are not going to believe the heavenly (spiritual) things." This is a serious problem and contributing cause to our frustrating Culture War, as students attend government (public) schools where an atheistic, evolutionary, billions-of-years view of history undergirds the entire education system. The students are taught—directly or indirectly—that the important Genesis narrative is not true, despite *not a hint* of long ages anywhere in the Bible.

Some churches, seminaries, and Christian leaders have compromised as well, as already mentioned, but Christians cannot surrender the Genesis record without surrendering the authority of Scripture (what Jesus was trying to teach Nicodemus). One could make a pretty good argument that all major Christian doctrine has its origin in the first eleven chapters of Genesis. Think about what happens when sharp inquisitive young minds, wondering about life's deep questions, and considering the claims of Christianity, are taught that the key singular events in these early chapters of the foundational book (Genesis), of the foundational book of books (the Bible), cannot be trusted and don't really mean what they say. Their most common (and justified) next question: "Well, then, what parts of the Bible can I trust?" An excellent question, indeed; distrust in the real history (the earthly things), then no trust in the spiritual (or heavenly) things.

It should be stressed at this point that without a real, actual, historic Adam and Eve, the Christian gospel, and even *the need for* a Savior and Redeemer, absolutely falls apart. The Christian "gospel," or good news, makes no sense without the "bad" news (a real Adam and Eve with disobedience and rebellion). I remember years ago a commentary by atheist Richard Bozarth from the *American Atheist*, and he seemed to understand this critical issue better than some Christian leaders:

> Christianity has fought, still fights, and will continue to fight science to the desperate end over **evolution**, because evolution destroys utterly and finally the very reason Jesus' earthly life was supposedly made necessary. Destroy Adam and Eve and the original sin, and in the rubble you will find the sorry remains of the Son of God. If Jesus was not the redeemer who died for our sins, and *this is what evolution means*, then Christianity is nothing.[171]

One thing obvious here (and in other writings by Mr. Bozarth): this guy truly understood that evolution is a direct attack on the foundation of the Gospel.

Karl Marx, along with Darwin and Wellhausen, also contributed to this move of Christianity away from divine revelation to a nebulous religion based on human reason. As hinted earlier, many other influential European intellectuals were involved in this ideological maelstrom, including evolutionists Charles Lyell, Thomas Huxley, Ernst Haeckel, Friedrich Nietzsche, Francis Galton, and Herbert Spencer. But Wellhausen's "higher criticism" and documentary hypothesis—with biblical truth and revelation denied— was a definitive turning point. Christianity began a quick decline in large part—once again—because of the false starting assumptions of Darwin and then Wellhausen, and religious liberalism won the day. As pointed out by author-historian Bill Cooper, Julius Wellhausen and his German *higher criticism* colleagues developed and designed this now disproven, but still somewhat fashionable, documentary hypothesis to be absolutely destructive of any impression that parts of the Old Testament— and particularly the first eleven chapters of Genesis—could be a reliable source of historical information.[172] Due to white lab coat intimidation and the belief that Lyell and Darwin had proven *millions of years and evolution,* the German scholar recognized the need to *fluidize* the text, because detailed Old Testament genealogy studies pointed to an approximate age of the earth of just 6,000 years or so. As time and events soon rounded the corner into the 20th century, the German rationalists' theories became entrenched in academia, then most seminaries, then churches and eventually more and more in the culture. They have unquestionably contributed mightily

---

[171] G. Richard Bozarth, *The Meaning of Evolution*, American Atheist, February 1978, p. 30.

[172] Bill Cooper, *After The Flood*, New Wine Press, West Sussex, England, 1995 (The early post-flood history of Europe traced back to Noah)

to *the mess we're in today*. The obvious message: Genesis is not real history; it's just a story, allegory or myth . . . and the Bible can't be trusted. Consequently, there arose an ungrounded exuberance and optimism that man and reason together could accomplish anything, especially through science, unfettered with biblical statutes or divine laws from God. So what was the cultural fruit? How did that all work out?

The bloodiest century in world history was soon unleashed upon mankind. Culture *is* indeed religion—**or *the lack* thereof**— externalized.

34

COUNTRY/RELIGION/CULTURAL FRUIT COMPARISON . . . NO CONTEST

Country	Foundational Religion/ Belief System	Cultural Fruit
Russia	Communism	Uncountable tens of millions of her own citizens murdered, starved to death, or worked to death in the 20th century Gulags (usually Siberian work prisons); no right to vote (or when allowed, virtually meaningless) or property ownership (until recently-on a very limited basis); depressed, unbalanced economy, frequent shortages of basic consumer goods, too dependent on oil export revenue; lack of freedom; suppression of religious freedom; pollution and poor environmental safeguards until recent years; 1986 Chernobyl nuclear disaster; national alcohol problem (more than 30% of all deaths in Russia in 2012 attributable to alcohol (WHO/OECD data)

China (mainland)	Communism	Uncountable tens of millions of her own citizens murdered, starved to death, imprisoned or placed in forced labor camps; massive food and consumer products shortages for decades; no right to vote; virtually no private property ownership; infanticide, forced abortions and childbearing restrictions; little religious freedom; serious pollution; worst air pollution in the world; new reports (2019; see World Magazine, Fox News reports) of internment camps in western China, with reports of murder for human organ harvesting (for medical transplants); government sanction of widespread international intellectual property theft
India	Hindu	Widespread poverty and illiteracy; limited religious freedom; unfair cultural caste system which is the basis for educational and job reservations, particularly against Dalit Christians; filth, unclean water sources, and extreme water and air pollution in the big cities (9 of the 10 most air-polluted in the world; bars and salons have even recently opened in some, for citizens to have a seat and buy and breathe concentrated oxygen through a mask for 15-30 minutes, for a fee); still tolerates female genital mutilation (FGM); still utilizes government-sanctioned (legal) gang rape for certain female crimes or Hindi law violations; limited female rights
Saudi Arabia	Islam	No religious freedom; no women's rights; polygamy; still tolerates female genital mutilation(FGM); few modern scientific, engineering, or academic achievements or contributions; rule by family oligarchy; no fair right to vote; finances Islamic terrorism; limb amputations as criminal penalties; disrespect and dislike of dogs
Bangladesh	Islam	Limited contribution to scientific, academic, engineering or industrial advances; one of the three most polluted countries in the world; limited female rights, tolerates female genital mutilation (FGM); poverty, low wages, and a partial ban on trade unions; widespread reports of human rights violations; government restrictions on free speech, media, and religion

Pakistan	Islam	Uneven income distribution, with at least 40% poverty rate; high illiteracy rate; limited women's rights; still tolerates female genital mutilation (FGM); Islamic terrorism; political instability; currently the most all-around polluted country in the world; has an unfair caste system; shortage of educational, hospital, and electricity-generating facilities; partial lack of religious freedom
Indonesia	Islam	Limited women's rights; still tolerates female genital mutilation (FGM); deforestation, soil erosion, and desertification from intense commercial logging; economic problems last twenty years, with government take-over of businesses (now owns over 140 enterprises); terrorism challenges; severe air pollution in major cities; poverty; water and sanitation challenges; partial religious freedom
Nigeria	Islam, especially in the north and southwestern parts (largest Muslim population in sub-Saharan Africa); Christianity more in the south; some traditional/tribal religions	Presently over 200 million people (seventh-ranked in the world), but ranks 140th (2018) per capita GDP. Limited women's rights, particularly in Islamic-dominated areas; 50% literacy rate among females, 61% for entire population; extensive and widespread poor, with 50%+ below poverty line, despite being Africa's leading oil exporter; considered most corrupt country in Africa, with widespread government corruption and misappropriation of public funds; high unemployment; serious malnutrition problem; hundreds of thousands of deaths and orphans thru HIV/AIDS; rampant deforestation, air pollution, soil abuse and degradation, and industrial and oil industry spillage poisons runoff, contributing to water pollution; heinous Boko Haram Islamic terrorist stronghold in north

Great Britain, France, Ireland, Holland, Italy, Belgium Germany, Switzerland, Austria, Scandinavia & Western Europe	Christianity	Large economic output, property ownership; freedom of religion; women's rights; great scientific achievements and discoveries, countries of origin for many of the founders of modern science; great works of art: sculpture, painting, architecture, music; country-wide modern sanitation and clean water; the right to vote and representative government; the rule of law; high standard of living; low illiteracy rate; low pollution
Australia and New Zealand	Christianity	Widespread voting rights and property ownership; women's rights; high standard of living; low pollution rate; freedom of religion; low illiteracy rate; academic, scientific, industrial, and economically advanced; low poverty; widespread modern sanitation and clean water availability; very high human development index; high per capita income. Ranks near the top, particularly New Zealand, for Social Progression, which includes such areas as Basic Human Needs, Foundations of Wellbeing, and the level of opportunity to its citizens.

United States	Christianity	Large economic opportunity and output; historically identified with individual liberty, freedom of speech and religion (the world's "crucible," or "melting pot"-formerly *"smelting* pot", a term describing the opportunities and fusion of different ethnicities, nationalities, and cultures); representative republic with voting rights; women's rights; widespread modern sanitation and clean water; large-scale conservation, fisheries, and wildlife management policies; property ownership; high levels of scientific, medical, engineering, architectural, agricultural, and high-tech/information technology research, development, and discoveries; origin of omnipresent world-changing technologies and products the last two centuries, including the assembly-line mass production of the automobile; the telegraph and telephone; the widespread use of the electric light, electricity, the modern television, and motion pictures; AC current and hydroelectric power plants; aircraft and air travel; spacecraft and advanced rocketry; lasers, personal computers, mobile phones, and the Global Positioning System; microwave ovens, cardiac defibrillators, hearing aids, traffic lights, and the Internet; as well as the all-important chocolate chip cookie, and the Ferris wheel, just to name a few.

It is unnecessary and space-consuming to list and categorize more countries; they and their parameters can all be found on the Internet or from various other sources. Many of these, with their distasteful cultural fruit of the last half century, should be well-known to any informed citizen, (e.g., terrorism-sponsoring Islamic Iran, communist Cuba and North Korea, or newly socialist Venezuela). Socialism was not listed specifically as a religion or worldview, because looking at *evidence-based* history, it is pragmatically synonymous with totalitarianism or communism, and has failed everywhere it's been tried, including the most recent attempt (in Bolivia, Nov., 2019). I remember almost a year ago when Nikki Haley, our UN Ambassador, issued this succinct socialism tweet: "3 million migrants, one million percent inflation, mass public protests-this is what Venezuelan socialism has done to its people." Further touching on socialism/ communism-relative to my borrowed maxim (*Culture is Religion Externalized*), I'll never forget something I observed on the television news while in school (around 1980). At least a hundred irate Russian women, obviously with no makeup on and bundled up in scarves and winter coats (all the same dark color), were protesting vociferously—actually borderline belligerently—in the snowy Moscow streets about the dearth of sanitary napkins and toilet paper (obviously more important than makeup). They were quite impressive, and flat-out *mad* (and a little bit scary); I'd wager even the KGB kept its distance. My comment at the time, to my roommate, went something like this: "This is amazing; here is a big, powerful country, a long-time nuclear power, former producer of great scientists, musicians, composers, outstanding writers and great literature (before communism), and they can't even provide basic necessities for their people?! You've got to be kiddin' me!"

On another related tangent, I used to help the Baptist Byelorussian ministry by contributing dental work on the kids from Belarus (Chernobyl nuclear disaster area). The kids were brought over for staggered six-week periods and hosted by gracious American Christian families and fed well, taken to doctors and dentists, perhaps taken swimming or to the beach, amusement parks or museums, and all kinds of good fun stuff. One particular soft-spoken, intelligent, beautiful girl about eleven years old that I worked on was really special, but the host family shared with me that she got very upset and threw a fit, after her first visit to their favorite supermarket to purchase groceries. I was puzzled at first, but then they explained that she thought it was all contrived—or essentially faked—and that the aisle after aisle in the supermarket loaded with foods, canned goods, a myriad of breads and rolls, packaged foods and candy, fresh vegetables, a plethora of fruits, meats, dairy products, and so much more that we experience every time we go (and take for granted), was just "set up" to impress her, and perhaps some of the other children. The variety and amount of food, beverages, household and other products to choose from was simply overwhelming, and totally unbelievable to her, especially after it was patiently explained that there were such stores all over the entire Charlotte, NC area, and countless other cities and towns across the country. Those wonderful kids were *wowed*-even stunned-at the American (and capitalistic) land of plenty.

Sticking with the socialism subject, and its seemingly resurgent popularity among college students and millenials, I had some head-shaking on a wonderful trip to Ireland this past summer. Almost every evening my wife, friends, and I were out strolling in the beautiful, historic and festive towns of the southwestern third of the country, in friendly charming places like Kinsale, Killarney, Dingle and Galway. Many pub and restaurant doors were open, the food and *stout* was great, colorful flagging, vendors, flowers and banners were everywhere, musicians were performing in the clean streets, thousands of people from various countries were mixing and having fun, and local business was booming. Everywhere I saw an Irish flag, there was an American flag right beside it, and it became amusingly apparent (and satisfying) to me that the most popular song over there was *Take Me Home, Country Roads*! It was so much fun . . . and no riff-raff, no terrorist worries. We were utterly enjoying ourselves one of those evenings, strolling through the streets and around the vendors and mingling with the cheerful, cosmopolitan crowd while listening to the music. Feelin' real fine and falling in love with the Emerald Isle, my reverie of that peaceful but upbeat atmosphere of freedom, free enterprise, good music, joy, and safety momentarily evaporated by an out-of-place side-walk table covered with pamphlets and literature. I immediately noticed the incongruous, somewhat dour countenance of the three young people at the table, and then I saw the sign: *Irish Socialist Party*. Along with a sign-up sheet, there were a few posters with statements that included the following (I actu-ally wrote them down): "Capitalism is destroying the nations," "Capitalism is the main cause of our problems," and "Help us to save our planet!"

I just shook my head, and bit my tongue. It was almost comical, their austere little set-up, ironically surrounded by the bustling effervescent Irish example of thriving capitalistic free enterprise. One young man actually reminded me of the young, intense communist-socialist political activist hus-band (played by Tom Courtenay) of beautiful Laura (played by Julie Christie) in the classic 1965 film, *Dr. Zhivago*. The socialism thing is not difficult to explain, even though it's failed everywhere it's been tried; they're getting it from the academic elites and professors, as well as your Bernie Sanders types, with support from the secular mainstream media. Alas, perhaps I need to be more tolerant, and patient. Even the cultural/political observer nonpareil Winston Churchill once commented: *If you're young and not liberal, you have no heart. If you're old and not conservative, you have no head.*

Moving on, it should be mentioned that the top 10 economies by per capita GDP are all from Western Europe, and the United States, with the exception of the tiny anomalies Qatar and Singapore. Capitalistic, west-ern-leaning Qatar is less than 4,500 square miles of desert peninsula and makes the list because of its ultra-rich oil and gas reserves (and is the only one without a dominant services industry), while multicultural Singapore, an island nano city-state and bastion of capitalism in southeast Asia, is a global financial center with a very dominant services sector (thus the high per capita GDPs).

The other 8 in the top 10 were all founded and established on the Judeo-Christian ethic and tradition. They all made mistakes, and had regrettable policies, events, and periods in their history. That always happens with *fallen*, or sinful, human beings no matter *what or who* the moral compass is. The point to consider objectively is which starting assumption, or worldview/reli-gion, leads generally to *more liberty, freedom, and hope*? Does one particular worldview/religion lead to not only more liberty but also a higher standard of living, better agriculture and ecosystem stewardship, more robust economic life and higher quality production, as well as better educational opportunities and more scientific, engineering, and medical advancements for the family of man? Does one really work best? If you were to go online and look at a world map of countries color-coded by *Human Development Index* (HDI), which is a statistical composite index of life expectancy, education, and per capita income indictors, the nations of North America, Western Europe, along with Australia and New Zealand would jump out at you-i.e., those with *a Judeo-Christian* founding and tradition. It is emphasized in this book, though, and obvious to any clear-eyed cultural observer, that these nations, including the United States, have unquestionably drifted significantly away from their Biblical moorings in the last century, especially since WWII. As Francis Schaeffer so clearly pointed out, the resultant negative cultural and societal parameters should not be surprising, and go hand-in-hand with the modern evolutionary view of final reality as *only material energy* (no God).

In the United States specifically, probably the most Judeo-Christian based nation in the last few centuries, it is truly remarkable what the fledgling, frontier, somewhat isolated young country morphed into in a mere 170 years or so, from 1790 to 1960, as it became the greatest economic powerhouse and most high-tech nation in the world, with liberty, freedom and opportunity, albeit with warts, growing pains, and the slavery issue (discussed earlier in this book). From the mid-late 1800s and thru most of the 20th century, more great inventions that impacted the entire world sprang forth from Christian-based America than any other country, and Americans were awarded more Nobel prizes in science and medicine than any other nation as well. Obviously it's Judeo-Christian Founding and culture did *not deter* scientific advancement; the evidence indicates just the *opposite*.

Things have changed in the last half-century, however, with the suppression of biblical authority and its replacement with evolutionary materialism, along with the removal of the Bible and prayer from public schools. The resultant cultural fruit includes the *un*surprising resultant sexual revolution and the rise of relative morality, diminished academic performances and standards, a dishonest mainstream media, and increased legislation and litigation (whenever *lack of trust* grows in society, legislation and litigation increase as well). The formerly united 50 states have now fallen behind other industrialized nations in various high school performance standards, overall scientific and engineering literacy, as well as Nobel prizes. Violence and mass shootings, gender-transgender dysphoria, stifling political correctness and uncivil discourse, a corrupt *deep state*, over 54 million abortions, skyrocketing indebtedness, rising suicide rates, no basis for law or Supreme Court rulings (all is arbitrary), and an ever-widening cultural chasm spreading from coast to coast with a new national angst has got a lot of Americans pondering . . . *what is going on*?

In consideration of pollution and the volatile issue of possible man-made climate change, and in light of my borrowed maxim, *Culture is Religion Externalized*, recent data shows that of the seventeen most polluted countries in the world, all are in Asia, ten have Islam as the dominant religion, two are Hindu-based, two favoring Buddhism/Confucianism, and three Communism (see WorldPopulation.com). The 20 countries with the unfortunate distinction of *lowest* gross domestic product (GDP) *per capita* (2017), are all in Africa or Asia, with the one exception being hard-to-fix little Haiti (Statista, Aug., 2019); *none* have a strong, historical Judeo-Christian foundation. The Human Freedom Index, calculated from a broad measure of human freedom, understood as the absence of coercive restraint, and looking at 79 key indicators (by the Cato Institute), shows the *lowest* levels of freedom exist in the Middle East and North Africa, sub-Saharan Africa, and South Asia (Islamic and Hindu dominated).

The *lowest* levels of women-specific freedoms are in the *exact same* areas. The data and real history indicate that any worldview or religion that is *not* Judeo-Christian inevitably suppresses or harms women. The data also shows that the *highest* levels of freedom are in North America (the U. S. and Canada), Western Europe, and Australia/New Zealand (again, those with a Judeo-Christian starting point). Am I saying that all of these countries at the top of the aforementioned undesirable lists, which are undergirded by Islam, Communism, Buddhism, Hinduism, or tribal religions . . . *are wrong*? Regrettably, words like *sinful, immoral*, and *wrong*, like many others in this 21st century, are now often defined differently by various people with different worldviews. Consider, though, my contention that *culture is religion externalized* . . . and you take it from there. The United States can't even begin to handle or accept all of the people that want (or try) to come here. Is it really true, then—as the left-leaning academics have claimed the last several decades—that America is bad, and responsible for most of the world's problems? What a joke. How many people do we observe spending their meager savings, and indeed sometimes risking their very lives, to get into Russia, India, China, Pakistan, Saudi Arabia, Bangladesh, Nigeria, Sudan, Iran, or Venezuela (just to name a few)? Don't misunderstand; I've met some fine people or have friends from five of those countries. Looking at the global picture, though, and honing in on nations with the most positive cultural fruit and parameters the last few centuries, is it legitimate and justified to more seriously consider their religion or god?

On reflection and comparison of these countries, in light of their dominant religion and resultant cultural fruit, the Devil is often in the details. The seemingly minor stuff—or details—can sometimes be revealing. It may be apparent to the reader by now that I love not only *real* science but also the great outdoors, and nature's panoply of plants and animals, including that ultra-special animal family *Canidae*. I have been around, raised, trained, loved and seemingly been loved by many wonderful and unforgettable house and hunting dogs.

I struggle with the Islamic religion for many reasons, but one of the irritating *details* is their disdain for dogs. I've got a *big problem* with that; such an undeserved attitude about dogs *alone* should make one skeptical about that religion. My experienced hiker friend Scott may be as well, as his devoted Texas Heeler (or shepherd), named Boone, helped save his life last year during an agonizing eleven hours on a remote North Carolina mountain, following a nasty encounter with a timber rattler. I strongly suspect that the notorious Islamic-state terrorist leader Abu Bakr al-Baghdadi[173] indeed had an acute, adrenaline-catalyzed, even super-heightened hatred of canines in his recent final moments. It was in a dark dusty tunnel in Syria, being chased and trapped in a surprise attack by a brave American military dog (named Conan) and our Special Forces warriors, before blowing up himself and three of his innocent children (Oct., 2019). Rest easy; the courageous canine was unfortunately wounded in the explosion, but is now well into recovery. I suspect al-Baghdadi never watched one of my favorite 1990s movies, *Must Love Dogs*. It's no big surprise if the prevailing Islamic perspective on dogs is not important to some of you readers. Think about it, though; perhaps it *should be*. Always pay attention to the little things.

---

[173] 49 people were killed and 53 more injured by 29-year-old Omar Mateen in the horrific Orlando nightclub shooting in 2016. The troubled shooter called 911 after commencing his attack, pledging allegiance to al-Baghdadi.

35

SOCIAL MEDIA, SILICON VALLEY, AND "SELFIE" CULTURE: MAJOR MISINFORMATION AND MANIPULATION

IT IS INTERESTING how historians and academic pundits classify or name different "ages" and "eras" throughout history. Some are appropriate and descriptive, while some are misleading and inaccurate-*not* based on reality or real historical and scientific evidence. Examples of the latter are the so-called Stone Age, the Iron Age, and the "Age of the Dinosaurs" —or the Mesozoic Era— just to name a few. The so-called "Iron Age," for example, is considered by long ages/evolution believers to have come after the imaginary "bronze" age, but the Bible mentions skilled craftsmen in both iron and brass even *before* the Flood (Genesis 4:22), 4,500-5,000 years ago. In Chapter 20, verse 24 in the ancient book of Job, some 4,000 years ago, the "iron weapon" and the "bow of steel" are mentioned. Others ages, chronologically listed here that most folks generally agree with, might include the Agrarian Age, the Industrial Age (with the concomitant Scientific Age), the Atomic Age (1945 on), and of course the recent and irrefutable Information Age (especially the last 30 years). But now we regrettably seem to have entered a new one: the *Misinformation* Age, which is a major catalyst for the writing of this book.

A huge part of that includes Amazon, Google and Facebook, their industry dominance, and Big Tech misinformation and deception which includes theft and misuse of private information, and the manipulation of algorithms to favor their liberal worldview and Democrat Party political candidates. They have a different perspective from most Americans on the foundational principle of personal choice and free will. In his fine book *World Without Mind-The Existential Threat of Big Tech*, Franklin Foer says "They hope to automate the choices, both large and small, that we make as we float through the day. It's their algorithms that suggest the news we read, the goods we buy, the path we travel, the friends we invite into our inner circle."[174] Censorship and sneaky manipulation of users abounds by these media giants of Silicon Valley, and in the case of Google and YouTube regular intervention from the top bosses as to which YouTube platforms and videos are acceptable.

Many users are dealing with smartphone and social networking addiction, with out-of-balance lives and negative self-image effects. It is interesting but not surprising that in this *selfie* age, self-esteem seems to be diminishing with complaints of *loneliness* and depression everywhere. It is even getting physically dangerous as well. It is estimated that several hundred people (mostly young adults) have lost their lives in recent years taking selfies- falling off cliffs and such (and those are just the ones reported); might be wise for folks to take their eyes off themselves a bit. More and more of young adult social media users are now attending therapy sessions or visiting psych doctors. Who would have thought that new and wonderful high-tech gadgetry and social networking use would lead to health problems?! Commenting on the influence of social media on the younger crowd, even Facebook founding president Sean Parker recently stated, "God only knows what it's doing to our children's brains." The unfortunate mindset among young adults today: life is all about *feelings* and emotions—not a good thing—social networking and the Facebook life are part of the problem. There they experience non-substantive gobbledygook, personal attacks and gossip, with all kinds of users talking mighty strong and opinionated, mouthin' off in a way they would never have the guts to say face-to-face as they criticize or slander some individual, or espouse some talking point they just picked up. It's gotten to the point where some lost souls literally live to respond on social media. They need to get a real life.

Even Super Bowl MVP quarterback for the Philadelphia Eagles and strong Christian Nick Foles chimed in on this issue over two years ago, around the time of the big game, and the myriad problems of social media and the Internet, with all of the temptation, negative, misinformation and pornography. For many young folks a thick skin is definitely required to deal with the sometimes profane and belligerent vitriol and non-civil discourse. Young women and men can be powerfully affected in a negative way by this, and the increasing numbers of subsequent suicides is stunning. (On further reflection, perhaps it should not be; overall suicide is up 33% in the last twenty years in this secular, confused society).

We are at a point now where schools should probably even have a required course in seventh or eighth grade on sensible social media use, with an important emphasis on the potential negative ramifications of their tweets, responses, posts, etc., which could be used against them in the future.

---

[174] Franklin Foer, *World Without End-The Existential Threat of Big Tech*, Penguin Press, New York, 2017, p. 3

I also lament the lost art of face-to-face conversation, partially related to the techie, selfie online culture. Increasing numbers of young people seem to prefer *not* to physically get together and socialize, exchange ideas, and—in a civil fashion—verbally and mentally spar; it's just too dern hard (what would have been unthinkable for my witty, articulate and riposte-savvy English hero, Winston Churchill). Smartphones, search engines, and Social Media have not only changed our attitudes on information and communication, but in many cases have promoted laziness and lethargy in research and truth detection—we don't want to work too hard, dig too deep, or spend much time—we want the information or answer easily (even if only half-true) and *right now*. The main members of *Big Tech* include Amazon, Facebook, Google, Apple and Twitter. Several of these companies are now more powerful than most countries of the world. Understandably, most Americans cannot and do not want to get rid of their smartphones or social networking, but let users beware!

More qualified authors than I have recently written sharply about the different aspects of this growing problem including their often hidden manipulation, bias, censorship and influence in so many areas. I used to be upset with liberal bias in mainstream media newspapers and television network news; still am, don't watch. Some fifteen or so years ago, when data showed that 75- 80% of Americans believed in God, research indicated at that same time that 90% of media elites were atheists. This figure surprised even me, and was a stunning illustration of the disconnect between the American people— especially those of us in flyover country—and the filters and purveyors of information, most of whom are associated with large urban areas. The problem has now shifted to a whole new level, with Big-Tech bias, algorithm manipulation, one-sided political leanings, and even strong-arm tactics. In his informative book, Mr. Foer cites such a bullying example by Amazon, who dominates Internet book sales and increasingly the book publishing market: "When sparring over terms with the publishing conglomerate Macmillan, it stripped the company's books of the buttons that allows customers to purchase them. In its dealings with Hachette (probably the largest book publisher in France-RJ) it delayed shipment of books. According to some who have sat across the table from Amazon, the company leaves no doubt that it will suppress a publisher's performance in its algorithms and eliminate its books from its emails if the company rejects its terms."[175] Amazon is unquestionably the behemoth of online book sales, with founder and *selfie* fan Jeff Bazos probably the richest man in the world.

Few people realize they have an even bigger income through their incredibly important, low-profile, esoteric security and surveillance business, involving facial-recognition software, Cloud storage, voice recognition

---

[175] Ibid, pp. 105-106

technology and who knows what high-tech else. I'm okay with security; it's the surveillance part that makes me nervous. They have become perhaps the most valuable company in the world in establishing omnipresent surveillance states, not only including stunning contracts with our own law enforcement and police departments, military, and CIA but many other countries as well, with minimal outside oversight and monitoring, and an increasingly obvious left-leaning political bias. The tech giant announced in the spring of 2019 that it is also getting into space exploration.

Amazon also likes to criticize and reject faith-based Christian groups. A wonderful Christian organization I've followed for decades, D. James Kennedy Ministries, is involved in a lawsuit against Amazon and the left-leaning Southern Poverty Law Center for false categorization of their organization and ministry as a "hate group," which is preposterous. The same *hate group* designation has been slapped on their distant cousin organization, the Alliance Defending Freedom (ADF), who was ejected out of the Amazon Smile Program.

We are at a point where Silicon Valley and Big Tech—with Amazon right near the top—has more influence and power than most countries and foreign governments. Some of the industry and business pundits recently claimed that Amazon has the most valuable "brand" in the world (over the other members of Big Tech, and everybody else), worth $315 billion. It's a new world out there, with privacy virtually out the window, and their influential surreptitious tentacles extending in ways most people cannot even imagine. And does anybody know how Amazon can do well over $200 billion worth of business last year (2018), with close to $11 billion of profit, and not pay *any* corporate income tax?! It would be interesting to know, I'm just not smart enough to figure it out. With a recent press release (early 2020), Amazon revealed its business up 20% in 2019 to a staggering $280 billion in revenue. This includes a net profit of 11.6 billion, or about 4%. Sounds small, but business experts say that is good and normal for corporations that operate on a 'big volume' business model. The press release indicates net product sales of $160 billion and net service sales were $120 billion. Book sales are now but a small part of their business.

Google rivals Amazon as the most powerful and influential company in the world, and there is a growing concern that it uses its ever-increasing influence for political reasons and leftist candidates, especially with its ubiquitous search engine. They now absolutely dominate the search industry, and their market share of the applications (*apps*) markets is well over 80%-a cause for concern to many industry observers. Facebook and Google combined have a large majority of the all-important online advertising market, which has negatively impacted in a significant way the advertising revenue of television, radio and other outlets. We're talking big power, big influence, and big money . . . bigger than big. American success stories for sure, and mega-profits are fine, but the real problem now is not only virtual monopolization and non-competition, but censorship and the smothering of ideas, especially those that don't coincide with the leftist Google party line. Censorship might be defined as *keeping something, or certain information, from people, to manipulate their thinking and decision-making*. More and more Americans are beginning to realize this, many millions are not.

This past year the Google CEO, Sandar Pichai, denied this as he stated before Congress "We don't manually intervene on any particular search result." This is doubtful after the leak of internal emails, more recently internal messages to Breitbart News, where employees of the tech behemoth mention such manual interventions to influence politically-related search results through Google's various vehicles, including YouTube. These revelations reveal discussions to manipulate search results to thwart various policies of President Donald Trump, including immigration and restrictions on Islamic terrorist-producing nation-states. Jessica Powell, who worked at Google, talks about Big Tech dangers in her book *The Big Disruption*, and claims that Silicon Valley thinks they're saving the world. Last year (March, 2019), it became apparent that Google was even trying to influence CPAC and shape conservative viewpoints, to "steer them a certain way . . . away from nationalism."

Folks, they are stealing information, censoring information, manipulating users and influencing worldviews to fall in line with their own. Not too long ago (June, 2019) Project Veritas revealed that a leaked document indicates that Google staffers consider PragerU, Jordan Peterson, and Ben Shapiro "Nazis," and suggested disabling their content from the *suggestion feature*. PragerU is a somewhat conservative media outlet named after its Jewish founder, Dennis Prager. Sharp columnist-commentator Ben Shapiro is Jewish as well. Apparently PragerU has already sued YouTube and Google over previous censorship applied to its videos, claiming the number of its targeted or restricted videos is over 100. Project Veritas shared that Google employee Liam Hopkins wrote "If we understand that PragerU, Jordan Peterson, and Ben Shapiro et al are Nazis using the dog whistles, why not go with Meredith's suggestion of disabling the suggestion feature?" After this umpteenth example of bias and possible censorship, spokesperson Jen Gennai, a director of *responsible innovation* in the Global Affairs division at Google, speciously responded: "Google has repeatedly been clear that it works to be a trustworthy source of information, without regard to political viewpoint. In fact, Google has no notion of political ideology in its rankings." Wow, I wonder if she made this statement with a straight face; an undercover video has footage of her discussing how to prevent President Trump from winning again. Project Veritas reveals her words from the undercover video:

> We all got screwed over in 2016, again it wasn't just us, it was, the people got screwed over, the news media got screwed over, like, everybody got screwed over so we've rapidly been like, happened there and how do we prevent it from happening again? We're also training our algorithms, like, if 2016 happened again, would we have, would the outcome be different?

Words, speech and language are not only very important and of course can be beautiful and inspirational, but beware of *anybody*—especially in positions of major influence—who repeatedly use the words "screw" and "like." Here is Ms. Gennai again, responding to Elizabeth Warren's recent call for a break-up of Google (which—unknowingly to Ms. Warren—just crashed her remote hopes, unless she does a 180, for the Democrat presidential nomination):

> . . . if Google is broken up, it can't prevent another "Trump situation." . . . all these smaller companies who don't have the same resources as we do will be charged with preventing the next Trump situation, it's like a small company cannot do that."

And some of you thought Google and Big Tech were wonderful, honest and neutral! Incidentally, Gennai's division is run by the Google executive (Kent Walker) who has a declared goal of destroying President Donald Trump's grass-roots populist-nationalist movement, making it a "blip" or "hiccup" in American history. (*Note: Thanks to Project Veritas and Breitbart News). This Big Tech crackdown on free speech, dissent, and conservative or Christian viewpoints is stunning. Google also just censored a Claremont Institute ad (they're an Upland, California based conservative think-tank)— another sneaky free speech attack—but Google claimed it "was just a mistake." I remember several years back humorist Dave Barry, when asked about future trends and culture changes, said something to the effect that we humans will eventually travel around in cars directed by Google, which will also likely control or own the roadways, and (this part made me laugh) "much of the solar system." I'm not laughing now.

But it goes back much further than that to the Obama years, when Google's executive chairman, Eric Schmidt, went all-in for Barack Obama in the 2012 presidential race, helping him secure victory. Franklin Foer writes:

His [Schmidt's] recruits scoured massive data sets to target voters with unprecedented precision. "On election night he was in our boiler room," says Obama campaign guru David Plouffe. These efforts made a difference. "Obama campaign veterans say that applying this rigor to their half-billion-dollar media budget made it 15 percent more efficient, saving tens of millions of dollars," Bloomberg reported after the campaign. This wasn't just a freelance effort. Google published a case study about the role it played in re-electing the president, titled "Obama for America uses Google Analytics to democratize rapid, data-driven decision making." The paper has attracted barely any attention, but contains brash claims about Google's centrality to the outcome of the election. . . .Google boasted about how its data helped the Obama campaign **shape the information** voters turned up when they wanted to verify claims made during debates, and as they mulled their choices in the days leading up to the election.[176]

In our video and visually-oriented culture, let's not forget Google subsidiary YouTube-it *is not* neutral, objective, or without an agenda. Back in early summer (2019), it was discovered that Google employees manually interfered with search terms, to promote particular videos on YouTube, while hiding others, in the week or so leading up to 2018's abortion referendum in Ireland. Most undecided voters do their online research on such an important issue in the final week or days before the big vote, which is exactly when Google surreptitiously altered the results of the related searches. These Irish citizens had no idea that they were *not* looking at the most popular and relevant selections, or results, but instead those selected for them by anonymous Google techies in Dublin. Google admitted earlier this year that it did "blacklist" certain videos. If one gets on YouTube with a video platform channel, and develops and attracts a lot of followers or subscribers, you make money-they'll send you a check (they make significant advertising revenue). But when you gain a following but don't fall in lockstep with their liberal worldview or politics, you may get zapped or *demonetized*. Most of the top Pro-Life video platforms have been forced out. It's not right, guys—they're forcing their beliefs, and denying opposing views, restricting conservative and Christian videos that don't fit their secular, leftist worldview.

Online searching is the foundation of Google's business, and most folks assume that it is neutral, or honest and objective. It was disconcerting when I realized this probably wasn't true and that some searches are biased. I was

---

[176] Ibid, pp. 123-124

disappointed in myself that I was surprised. Are you parents out there at all concerned that your kids/students are mostly relying on Google education apps? As my apps-savvy friend explained, it involves manipulation of algorithms to favor certain viewpoints, worldviews, or political candidates, as was done with Hillary Clinton (those sneaky Google nerd-punks). Google is not politically neutral, as also already alluded to relative to the Obama presidential win in 2012. In a recent interview (Aug. 5, 2019 on *Fox and Friends*) on Fox News, Kevin Cernekee, who was fired for discussing with the Google hierarchy bias and mistreatment against conservative Google employees, said that some of the Google executives or managers actually *cried* when Donald Trump was elected. Mr. Cernekee even took his mistreatment relative to this issue to the labor board, but to no avail; he was terminated. He also cited regular censorship of conservative political ads, as well as similar actions by YouTube, and such bias at virtually every level at Google with the reality that employees will be targeted if they buck the leftist party line. Despite increasing numbers of similar complaints, Mr. Cernekee frustratingly admitted that no one has ever stepped in to do anything about workplace bias at Google.

Concerning the cozy Google-Democrat Party-Obama/connection, author Franklin Foer reminds us that **Google's head lobbyist visited the White House 128 times**. These apparently paid off, as with obviously more than just serendipity, "Google managed to overcome the recommendations of staffers on the Federal Trade Commission who found Google's monopolistic machinations worthy of a lawsuit."[177] I hope regular users out there also realize now that the *most searched* at the top of the list or page is a joke. It's so frustrating how cultural issues that seem like they're everywhere are being rammed down our throats. I'm talking about issues that perhaps many of your friends and the majority of Americans do *not* support or agree with-whether its gender fluidity, homosexual marriage, full-term abortion, socialism, open borders or something like the obnoxious and stifling coverage of Maxine Watters in late 2018, with her belligerent anti-American rhetoric. Big Tech shares much of the blame.

It is hard for us seasoned citizens to believe Google is twenty years old now, with widespread worldwide influence, probably knows too much about us, and unquestionably is one of the most powerful companies in the world. It is now known that this accumulated information becomes available, or for sale, to other entities, with your privacy and best interests of no concern. The wisdom and foresight of Christian warrior Francis Schaeffer still amazes me, as he seriously discussed in the 1970s—well before the ubiquitous presence and use of personal computers and cell phones—the potential problems of extensive information storage, computer dependence, and

---

[177] Ibid, p. 128

the wholesale monitoring of communications and electromagnetic transmissions, that "leaves no place to hide and little room for any privacy."[178]

Are you bothered that we may have reached a point where folks now need protection *from* Google? As economist and writer Luigi Zingales said in a 2017 Hillsdale College Free Market Forum: "Google and Facebook now know more about us than our spouses or closest friends—and sometimes even more than we know about ourselves. They can predict what we're going to do, how we're going to vote, and what products we're going to buy. And they use the best minds in the world to manipulate our decision . . ."[179]

Google and Silicon Valley have adopted the diversity/victimology gospel also, and caution is advised if bucking the party line. Young but sharp James Damore was fired by Google in late 2017 for his memo hinting that different inherent abilities and predilections in females and males may be the cause of the tech giant's unreached 50%/50% goal of gender employee balance. Up until just a few years ago, every sensible citizen would agree that only an idiot would deny that women and men—females and males—are wired differently, and that a plethora of evidence exists pointing to differences—not in intelligence—but in aptitudes and interests. That view is now a big *No-No* at secular institutions and large, left-leaning group-think corporations in this puzzling age of gender pseudointerchangeability. As of this writing, Mr. Damore and David Gudeman are now suing Google for discrimination against conservative male employees. If you're a white male considering pursuit of an entry-level computer/electrical engineering job at Google or certain other Big Tech firms, you better think twice. Sex (gender you were born with), skin color, and perhaps politically correct gender herd-think conformity often trumps ability.

Putting trust in left-leaning tech giant Apple is questionable as well. Just recently (end of July, 2019), a whistleblower revealed that Apple contractors listen to some SIRI conversations of users, including medical information, drug deals, and intimate conversations. SIRI is a virtual assistant in some of Apple's systems that uses a natural-language user interface to answer questions and make recommendations. Beware of Big Tech misuse of your private information and spying.

Some observers think Facebook is the most powerful and influential Big Tech media giant. Many folks, businesses, non-lefty politicians and even other media organizations have accused Facebook of violating privacy the last few years, as well as manipulating how you think. The *Information for*

[178] Francis A. Schaeffer, *How Should We Then Live?* Crossway Books, Wheaton, Illinois, 50th L'Abri Anniversary Edition, 2005, p. 244

[179] Luigi Zingales, *Should We Regulate Big Tech?* From *Imprimus*, Hillsdale College, Vol. 47, #11 (Nov., 2018)

Sale business has become a volatile issue with Facebook; they have apparently given *intrusive* access (by outside entities) to multiple millions of users, without the user's permission, of course, with big money made. Wow, shared user data, without permission. It's that subtle, sneaky dishonesty of Big Tech that irks me. Or, maybe I've got it all wrong, as Facebook attorney Orin Snyder recently claimed in court that "There **is no invasion of privacy at all, because there is no privacy**." Back to author Franklin Foer: "With even the gentlest caress of the metaphorical dial, Facebook changes what its users see and read", and then adds that they are "constantly tinkering with how its users view the world." The evidence indicates this includes censorship of posts by conservative and/or Christian commentators; just ask the charming African-American female commentators "Diamonds and Silk." On another front, Breitbart News reported months ago that Facebook has allegedly refused to allow the upcoming Roe v Wade film to run advertisements on its platform under the new "issues of national security" rule, which makes no sense at all (could it be because of their far-left, pro-abortion politics and worldview?!).

Facebook just announced (summer, 2019) somewhat unbelievable plans to create an alternative cryptocurrency (called Libra) financial system in cahoots with 27 corporate partners, with many more to come by the time the system opens in the next year or two if they can pull it off. A bold and questionable comment by former president of PayPal David Marcus, who now directs a major part of Facebook's technology research: "It feels like it is time for a better system." Oh, really? The system we have is working fine, the U. S. economy is absolutely booming and the envy of the world, and now one of the most powerful companies on the globe that does not even protect the private information of its users, and surreptitiously tries to influence how we think, what we buy, and who we vote for expects us to confidently buy into their own financial system. Patriots and potential users might want to go slow on this one; Facebook just got slapped with a $5 billion fine (late July, 2019 . . . just a slap on the wrist), as a result of an investigation into its mishandling of user data. The Federal Trade Commission claims Facebook has agreed to accept more oversight, but in reality no specific effective safeguards or rules have been implemented to protect user data and privacy in the near future. We've got a serious, complex, hard-to-handle problem in the former land of liberty.

Twitter is guilty as well, *shadow banning* prominent conservatives on its ubiquitous social media platform, altering their algorithms, denying their access and technologically censoring certain Republicans, like then Congresswoman Marsha Blackburn of Tennessee who last year was barred from advertising her Senate campaign launch video on their service. Twitter was ticked because of Ms. Blackburn's statement, "I fought Planned Parenthood, and we stopped the sale of baby body parts." What decent person could be against that? Ms. Blackburn won her Senate race, to the chagrin of Big Tech. They seem to never suppress liberal voices.

Republican Congressman Devin Nunes of California, who stood strong against the political left, their water carriers in the mainstream media, and the Hillary Clinton bought—and—paid—for fake Russian dossier story of Trump-Russian collusion, has just filed a multimillion-dollar lawsuit against Twitter for anti-conservative bias. He also accuses them of shadow banning conservatives and says he has been defamed and slandered hundreds, even thousands of times, and that the Twitter folks are "content developers." It will be interesting to see what happens with this—Twitter has plenty of money, influence and good lawyers—but Nunes has plenty of facts, smarts and steel backbone. There is no big surprise here. Not long ago Twitter's president admitted to their left-leaning proclivities, and it unquestionably has become *the outlet* or platform for the vocal Democrat party base.

When Twitter takes an individual's profile and followers, and makes their posts less accessible to those followers, it obviously reduces reach and exposure. In such cases the purpose and effect is to minimize followers. They want to diminish influence, especially if of a conservative bent.

The deception, bias, and hypocrisy are astounding. The liberal left has a now un-camouflaged double standard on free speech: censorship on Christians, conservatives, and the Right; with no censorship or suppression of the Left. There are some great individuals working in Big Tech, but studies by Barna research and others in recent years show that 80-90% of media and high-tech elites and high-level managers are atheists. It might be wise to remember that real experience and history show that people without a legitimate divine compass almost always have an agenda; they often don't care what the facts are or what the data shows. They don't necessarily care about decisions, interests, programs or strategies that really work— or are best—for the majority of the folks. More than a few of our Founders some 240 years ago proclaimed that the new American form of government—that unique "experiment in liberty"—would not work without a virtuous people. In this Misinformation Age, have the powerful purveyors of information lost their virtue? We've got a problem, folks.

What's to be done? An accumulating body of evidence even indicates strong connections and influence over the Democrat Party. We're essentially dealing with not only powerful political bias, but mega-dominance, non-competition and monopolies in social media, smartphones, online advertising and the applications markets—huge companies with tremendous influence and power. Dr. Zingales, in his commentary on the subject of the growing need for some sort of Big Tech regulation, refers to Haskel and Westlake's point in *Capitalism without Capital* where they highlight in the burgeoning tech sector not so much the usual tangible assets, but *intangible* assets like research and development, marketing and software. The reality and characteristics of such an intangible system, however, lead to higher market concentration and *less* competition. Dr. Zingales points out that the sheer magnitude and detail of accumulated and stored data is incomparable and unimaginable, and actually **inhibits competition.** He shares that the Google founders *even admit* that their amassed history of searches creates a huge barrier to new entrants (competition).

Not too many decades ago the monolithic Bell Telephone System had some 21 companies under the huge holding company and long-distance carrier AT&T, Southern Bell being the big one in the southeast, including here in North Carolina. AT&T also had the important subsidiaries Bell Canada, Western Electric (which included the world-famous Bell Labs), and three others. The point: we're talking dominating major monopoly here. On December 31, 1983, the system was divided into independent companies by a U. S. Justice Department mandate. The government stepped in and broke things up—whether it was done the best way possible is probably debatable. Among other things, every patent they got eventually had to be shared.

We now need some type of oversight or government intervention with Big Tech. It is not going to limit or fix itself. Such intervention occurred with AT&T, IBM, Microsoft and others and it worked, leading to not only a more level playing field, curbed influence and more competition, but new products, businesses and economic growth. Silicon Valley and Big Tech monopolies are so powerful now that if some action is not taken, their power and unwanted influence may enable them to accomplish their goal of molding humanity into their desired image of it.[180] Franklin Foer adds, "They believe that they have the opportunity to complete the long merger between man and machine—to redirect the trajectory of *human evolution*."[181] If intervention is not taken, conservatives, Republicans and even some independents may not win any major elections before long, and more and more young people may not control how they think. This is not an exaggerated overstatement.

Smartphones and gadgetry are great, but beware . . . there are no citizen secrets anymore, but Big Tech hidden agendas everywhere. They are uprooting the fences that protected privacy and personal choice. As a techie friend recently pointed out to me, "Caution, digital is forever." It is indisputable that they want to control *how you think*. And when I first heard about *no-mobile phone phobia* recently, what some are calling nomophobia, used for those individuals who develop anxiety and fear of forgetting, losing or being without

---

[180] Ibid, Foer, p. 2

[181] Ibid, p. 2

their smartphone, I just shook my head. The phone has become that much a part of their being, their very self?! Along with the new gender dysphoria issue, suppression of free speech, the rise of identity politics on college campuses, and the now omnipresent politically correct thought police, is it any wonder our young people have an identity crisis? While I'm not recommending profanity or curse words, have any of you parents, baby boomers, bosses, or older millenials ever just wanted to shout, "Put down your damn phone"!?

I think of us as spiritual beings with a human problem, more than human beings with a spiritual problem. Nourish your spirit—your *real* self; put the gadgetry aside occasionally. Several years back a Stanford study showed that the best all-around exercise for seasoned citizens was "walking in a rural environment." You millennials and young people need to try it, too . . . but turn your phone off . . . and don't forget to smell the flowers.

36

TROUBLED YOUNG MEN AND AMERICA'S MASS SHOOTINGS—WHY?

ON APRIL 20ᵀᴴ, 1999, teenagers Eric Harris and Dylan Klebold murdered 13 people and wounded 21 others by gunfire at Columbine High School in Littleton, Colorado. They then committed suicide. At that time, it was the worst high school shooting in U. S. history, and shocked the nation. Three months later, Mark Barton shot nine people to death in some Atlanta office buildings after he had already killed his wife and two children. This troubled stock market day-trader committed suicide several hours later. That terrible tragedy is one of well *over fifty* mass shootings in the U. S. which involved the deaths of five or more people, with many more wounded, in the last twenty years, not counting many other shootings with fewer casualties. Many of the shooting sites were cupcake or *soft* targets like churches. Others included airports, malls, businesses, theatres and night clubs, but many were at schools or colleges-in the shooter's warped mind: young, defenseless, and vulnerable. We have always had bad guys, criminals and troubled souls, and evil is indeed a real thing in this fallen world, but this is a new American phenomenon. An entirely new and embarrassing niche retail market with exploding sales has even appeared within the last two years: bulletproof backpacks for youth and students! Other countries now associate disturbed young men, their mass shootings and explosive violence with American culture. This is a far cry from the former stalwart breed that settled a raw new land of opportunity and in 200 years established one of the greatest countries in world history, while preserving freedom and hope by helping to win two world wars. What is going on?!

After each recent shooting, including the heinous Parkland, Florida high school shooting of February 14ᵗʰ, 2018 that left 17 dead and 17 wounded, as well as the early November horrific massacre in Thousand Oaks, Ca., where a troubled former Marine executed 12 Americans at the Borderline

Bar and Grill and then apparently turned the gun on himself, the "blame game" starts in earnest. The Democrats go crazy, blaming the Republicans; the NRA is at fault; and the violent video-game culture again comes under attack. 21-year-old Patrick Crusius recently opened fire on the morning of Aug. 3, 2019 and massacred 22 innocent people while wounding 24 more at an El Paso Walmart store. Writer Robert Evans, in a recent commentary relative to this young assassin's favorite message board on the Internet, discusses the apparent "gamification" of terror. Mr. Evans claims "mass shooters are translating their Internet personas into the real world." Shameless Democrat presidential hopefuls Beto O'Rourke, Corey Booker, Bernie Sanders, and Kamala Harris were among those who embarrassingly raced to exploit these newest massacres and blame the President for the shootings, with stunning, unjustified hate-filled vitriol unmatched in the past decade. Other points of blame: Hollywood, with their focus on guns, violence and killing; pop and rap music; the negative and dark side of social networking and the Internet; the removal of God, prayer, and Christian values from public schools; the breakdown of the family and the lack of strong parenting, and the blame game goes on and on.

One important part of the underlying cause of this terrible problem, which is rarely mentioned: the bedrock foundational axiom of public education—Darwinian evolution—also now known as Naturalism, or in recent years simply "natural selection." Many of you millennials and other readers may not have realized prior to this book that evolution, or what I refer to as *evolutionism*, actually undergirds and infiltrates our entire educational system, and has for many decades, initially catalyzed by Harvard University in the late 1800's and then spearheaded by John Dewey and other humanists in the early 20ᵗʰ century. All of this happened fast despite the lack of supporting scientific evidence, which is covered in other parts of this book. If you doubt the claim of the overarching importance of secular evolution, especially in education and government policy, please note the following commentaries starting with Sir Julian Huxley, one of the leading evolutionists of the 20ᵗʰ century:

The concept of evolution was soon extended into other than biological fields. Inorganic subjects such as the life-history of stars and the formation of the chemical elements on the one hand, and on the other hand subjects like linguistics, social anthropology, and comparative law and religion, began to be studied from an evolutionary angle, until today we are enabled to see evolution as **a universal and all-pervading process**[182].

Huxley had an important and influential position heading the increasingly secular United Nations' Educational, Scientific and Cultural Organization (UNESCO), and had directed its foundational focus and philosophy in its early years after World War II. Among his many significant comments relative to his atheistic evolutionary worldview was the following:

> It is essential for UNESCO to adopt an evolutionary approach . . . the general philosophy of UNESCO should, it seems, be a scientific world humanism, global in extent and evolutionary in background . . . Thus the struggle for existence that underlies natural selection is increasingly replaced by conscious selection, a struggle between ideas and values in consciousness.[183]

Rene Dubos, a French-born American microbiologist, environmentalist, humanist and Pulitzer Prize winner for his book *So Human An Animal*, said this in the mid-1960's:

> Most enlightened persons now accept as a fact that **everything in the cosmos**—from heavenly bodies to human beings—has developed and **continues to develop through evolutionary processes**. The great religions of the West have come to accept a historical view of creation. Evolutionary concepts are applied also to social institutions and to the arts. Indeed, most political parties, as well as schools of theology, sociology, history, or arts, teach these concepts and make them the basis of their doctrines.[184]

---

[182] Julian Huxley, "Evolution and Genetics," chapter 8, p. 272, in *What is Science?* (J. R. Newman, ed., Simon and Schuster, New York, 1955)

[183] Julian Huxley, "A New World Vision," *The Humanist* 39, p. 35-36 (Mar./Apr. 1979).

[184] Rene Dubos, "Humanistic Biology," *American Scientist* 53, p. 6. (March, 1965).

I hope you readers bear with me here, especially you young men—virtually all of the mass shooters were young men—because it is very important to stress the all-encompassing worldview and influence of evolution. Molecular biologist Michael Denton wrote the excellent book, *Evolution: A Theory in Crisis* in 1985, and also in 2016 a follow-up book, *Evolution: A Theory Still in Crisis*. Talking about *the influence* and acceptance of Darwin's idea in his first book, he said:

> The idea has come to touch *every aspect* of modern thought; and no other theory in recent times has done more to mold the way we view ourselves and our relationship to the world around us. The acceptance of the idea one hundred years ago initiated an intellectual revolution ***more significant and far-reaching*** **than even the Copernican and Newtonian revolutions** in the sixteenth and seventeenth centuries.[185]

A final quote from a now-deceased but very influential 20th-century evolutionist, Theodosius Dobzhansky, originally from Russia and who was on the faculties at Columbia and Stanford, among others:

> Evolution comprises all the stages of the development of the universe: the cosmic, biological, and human or cultural developments. Attempts to restrict the concept of evolution to biology are gratuitous. Life is a product of the evolution of inorganic nature, and man is a product of the evolution of life.[186]

This audacious and arrogant statement by Dobzhansky, one of the white lab coat intellectual elites in the 1960s- one of *the experts*- as if he is God, is just flabbergasting. No evidence, whatsoever . . . a process that **has never been observed in the present world or in the fossil record** . . . and goes against the known laws of science, including the universal laws of conservation and decay (the 1st and 2nd Laws of Thermodynamics), and the Law of Biogenesis. It is truly profound. It reminds me of the Apostle

---

[185] Michael Denton, *Evolution: A Theory in Crisis*, Adler and Adler Pub's, Inc., Chevy Chase, Md. 1985, p.15

[186] Theodosius Dobzhansky, "Changing Man," *Science* 155, p. 409. (Jan. 27, 1967).

Paul's closing plea to Timothy: "keep that which is committed to thy trust, avoiding profane and vain babblings, and ***oppositions of science falsely so called***" (1 Tim. 6:20). Indeed, evolution has become a religion to these people; most even admit it.

Eric Harris, one of the teenage Columbine killers (April, 1999), had written on his website, "You know what I love ??? Natural SELECTION! It's the best thing that ever happened on earth. Getting rid of all the stupid and weak organisms." **Evolutionary dogma unquestionably played a significant role here.** He wore a dark shirt emblazoned with **Natural Selection** on it, and the two troubled students selected the Columbine massacre day**, April 20**[th], in honor of the birthday of **their evolution-driven hero—Adolph Hitler**—himself a big proponent of *survival of the fittest*. Sadly, this is quite revealing, and makes one wonder about a public education system that is based on Darwinian evolution and humanism. This is very different from the 18th and 19th centuries. God, the teaching of creation and intelligent design, prayer, and Bible readings are of course banned in virtually all secular/public schools today. The Bible and prayer were officially taken out of the public school system in two successive years in the early 1960's by the left-leaning Supreme Court. This initial aggressive assault on public religious expression was facilitated and championed by Chief Justice Earl Warren, who was committed to redirecting the Constitution into a change-able, non-grounded evolutionary document.

Interestingly, right after this removal of the Bible and prayer from public schools, SAT scores **dropped 17 years in a row**—this had never happened before—they always had pretty much stayed the same. The somewhat sudden appearance and growth of Christian home-schools and private Christian schools finally stopped the decline. Remember my borrowed maxim: *culture is religion* (or the *lack* of religion) *externalized*. The Darwinian evolution was and is taught as fact, with a history of many documented fraudulent claims and illustrations (see Haeckel's embryos, the *geologic column*, Darwin's *tree of life*, as well as notorious Piltdown Man and Nebraska Man, just to name a few) in the most frequently utilized biology textbooks and museums. If kids are taught that they are just an accident, just an animal, just randomly evolved from algae and apes (for which there is absolutely **zero** scientific evidence), and that there is no God, *how can they have a purpose?* How can their life legitimately have any meaning or value at all? Part of Darwin's legacy: *Your morality can be whatever you desire it to be.* Consider today's dominant education foundation—evolutionary materialism—and look around you in the current culture: No God, no hope, mass shootings, increased suicides, and sexual perversion everywhere. A common mindset: *What does it matter? Why shouldn't I do what I want to do?!* Believe it or not, deep down young people want guidance, discipline, structure, and rules . . . along with hope. With the materialistic evolutionary worldview, where do the guidelines—the moral absolutes—come from? The honest answer: *nowhere*— no moral compass . . . just set your own . . . have at it with gusto. Remember though, when all of the people do what's right in their own eyes, just chaos. Do you doubt? Check history, check the fruit, and check todays' culture.

Nicholas Cruz, the troubled 19-year old shooter who on Valentine's Day, 2018, killed 17 and injured or wounded 17 more at a high school in Parkland, Florida was previously apprehended a half dozen times, and was visited some 30 or more additional times by local law enforcement for violence and questionable behavior. He was obviously a problematic and volatile young man-there were multiple warnings and signs. Why was action not taken? This was the same with the Korean-American Virginia Tech shooter in 2007 that left 32 dead and some 23 or so wounded or injured. A part of the answer is simply the schools and law enforcement acting more seriously and aggressively—political correctness aside—on warnings and red flags; it would have likely prevented some shootings. The troubled young Dayton assassin who had made online references to Satan and had a known history of being fascinated with violence, Connor Betts, who killed ten (including *his sister* and himself) and wounded 27 more on August 4th, 2019, had at one time made a hit list of schoolmates from high school that he *wanted to kill or rape*. This was discovered, but why was no action taken, like at least adding his name to a database so it would be more difficult to purchase a gun?

It seems that lax high school, university and government policy (state, county, and city too-not just federal), political correctness and *tolerance* are all affecting law enforcement and putting students and others at increased risk. It is similar to the politically-motivated ridiculous *rules of engagement* implemented on the battlefield by Beltway elites (people you'd never want beside you in a foxhole), especially in the previous administration, that have hamstrung our dedicated military warriors, complicated operations and even led to more American casualties. Oh, the politicians and swamp dwellers! And consider the following: In the horrific 9/11 Islamic terrorist attacks at the New York twin towers, the Pentagon in Washington, D. C., and the thwarted attack by some heroic Americans who gave their lives in the Pennsylvania country-side in 2001, nearly 3,000 Americans lost their lives and thousands of families were disrupted or destroyed, along with unknown billions in property and business losses. The *Deep State* is real, and the *D.C. Swamp* is real. **Not one person** in Congress, the FBI, the CIA, the NSA or law enforcement was held accountable, punished, lost their job, or went to prison. That reality is still absolutely, stunningly, irritatingly unbelievable!

Back to the troubled young men, the heart of the problem is a problem of the heart, but dietary and nutritional deficiencies, as well as powerful antidepressant prescription drugs, play a significant part in this national tragedy. Clinical depression or anxiety, which seemed to be **a common denominator** in these men, is usually related to a Vitamin B deficiency— more accurately called **a B-*Complex* deficiency.** In such a nutritional vitamin *complex,* there are many, many components; you need them *all* for complete function and good health. Included are Vitamin B-1 up through B-12, combined with trace minerals, trace mineral activators, co-factors and more. The point is, the human body *needs them all*, not just a part or *fraction* of them, which is what you get with the typical store bought B-vitamins and others. Such a B-Complex deficiency is unfortunately common in America, almost **epidemic in those young people** with the preferred dietary mishmash of sugar, soft drinks, processed and junk foods (or chemicalized *non-foods*), fast food and usually no pure water drinking. Regular, daily consumption of carbonated beverages will also inhibit B-vitamin absorption (as well as some other important minerals, like calcium) because of the carbonation. Many Americans simply **don't eat whole foods** any more, especially those rich in B-vitamins which include organ and lean meats, organic eggs, beets, green leafy vegetables, fish and seafood, brewer's yeast, real complete whole grains, and whole milk, just to name a few. This **is a *huge* part** of the violence and mass shootings problems with America's young men: the nutritionally-deficient garbage, chemicalized non-foods and junk they regularly consume and live off of. It is for the most part unknown, overlooked, or denied by the experts. We ignore this factor at our own peril.

Also contributing to depression and mental health/psychological issues is a dietary lack of *good fats and oils*, sometimes referred to as Vitamin F or *essential* fatty acids, as well as the continued ridiculous promotion of the *low-fat* diet by conventional medicine and many nutritionists. Readers, you cannot be healthy, with long-term mental sharpness and good brain function, without *good* fats in your diet. Think about the gray matter of your brain. It is in large part fat and cholesterol. One possible serious side effect doctors and drug manufacturers do not like to talk about is early memory loss which can lead eventually to dementia or Alzheimer's, from taking a blood pressure drug in combination with a statin (cholesterol-lowering) drug, especially if started relatively early in the patient's life. Alzheimer's seemed to suddenly explode on the national scene after two or more decades of unfounded "low-fat" diet promotion by the FDA, nutrition experts, and physicians along with a significant increase in prescription drug use. Some sharp alternative doctors therefore consider Alzheimer's partially an *iatrogenic* (doctor/hospital/drug-induced) disease; the jury's still out on that one. The point here: critical good omega-3 fats and cold-water fish oils tend to be **missing in the teenage diet. This deficiency, combined with a lack of B-vitamin complex, may contribute to depression, unjustified or vague fears, and emotional problems that can become catastrophic if left unchecked, progressing to anxiety, irritability, or paranoia right on to outright anger, hostile behavior and rage**. If you have a child or teenager that seems to have a strange recurring feeling *that something bad is about to happen*, be on high alert.

Mainstream doctors and media pundits may be unaware of this or outright deny it. I do have a little patience and sympathy with them because most studies indicating that vitamin supplements are unnecessary, worthless, or make no difference are performed with **synthetic, *fractionated* store-bought vitamins** that are manufactured with heat and pressure. They are made for pennies, sold for big bucks, for the most part **do not work** and may even be harmful. Americans waste untold millions of dollars on them. They are a far cry from powerful organic phytonutrient complexes and whole-food concentrates, with the life force intact, *all* of the components and *fractions* present in the vitamin complex, and formulated with special equipment. Most medical *experts* and well-known but uninformed commentators still deny or minimize this diet/vitamin deficiency/mental illness connection. But it is irrefutable to dedicated alternative medicine and holistic docs out there dealing with this.

What is the mainstream medical remedy for these troubled young men? Just load them up with drugs! These harmful and overprescribed antidepressants and powerful, psychotrophic drugs like Ritalin and Prozac **can mask important nutritional deficiencies and all have bad side effects, which ironically include mania, hallucinations, rage and suicide!** Most of the young killers were taking one of these drugs, or recently had been, and were presently or recently under psychiatric care.

Not recognizing the brain/nutrition connection, or the ramifications of drugging young people for life, is unconscionable. We have millions of kids with *starving brains*. Get them off the junk, sugar and non-foods; feed them real food. Get your kids to eat the brain foods I mentioned earlier, cut out the fast foods, sugar and carbonated beverages, stay away from packaged processed nonfoods (including most store-bought cold breakfast cereals), and make pure water the main drink. Try to contact a good naturopath, alternative or holistic doctor or a sharp nutrition-minded chiropractor for vitamin and supplement advice. Don't buy them at the store; these are usually isolated chemicals, or *fractions*, of the particular vitamin complex. When you take, say, a typical store-bought fractionated vitamin, your body recognizes internally there are missing components— it **is *incomplete*** (fractionated)—so it **will often *rob* areas of your body**

internally for those missing components, creating additional deficiencies. The phrase *nature abhors a vacuum* applies here; your amazingly designed system can tell there are missing but needed components or minerals, and it thus searches for them within, stressing your system, often making the deficiencies **even worse**. Only whole foods or whole food organic food complexes can provide them.

Your health and your kid's health (at least for a while, until they get older), are *your* responsibility-not that of the FDA , AMA, NIH or Dr. Fauci, the WHO, family doctor, or Google. Again: drink pure water, stay away from sugar and carbonated drinks, consume a whole-foods diet with a variety of fruits and vegetables, including some raw veggies—but *not* drowned in unhealthy store-bought *ranch* dressing. Avoid processed food, drastically cut back on wheat and bread, use a good sea salt, consume good fats and omega-3 fatty acids, stay off worthless cholesterol-lowering statin drugs and get some regular exercise. Avoid all drugs if possible. With ADD and behavior-problem kids whose parents strongly commit to this nutritional emphasis, a happy family can often emerge again. With the elimination of carbonated and sugary drinks, processed and fast food, while converting to a whole foods diet (like the Mediterranean diet) and a serious **switch to pure water**, while adding a comprehensive organic B-complex supplement, the ***results can be profound***.

The mass shootings phenomenon is a huge problem in what is now almost post-Christian America. Let's look at the whole picture. Think about the loneliness, the decline in face-to-face interaction, the small compromises, the bullying, the imagined slights, nutritionally-starved brains, and the academic and media suppression of men. Throw in violence-saturated video games, the dark side of internet use and exposure along with the negative side of social networking. Then there's confusing gender confusion, and lukewarm or non-existent parents . . . all undergirded by an evolution-based no- hope worldview taught in the school system with God, the Bible, purpose and prayer removed. *Junk* and non-foods have taken over young men's diets; *junk* material things and non-values have taken over their minds and hearts. **We seem to have become a nation filled with detached, desensitized, expressionless young people.** It has been reported that many of the killers seemed to exhibit a stoic, terminator-like expression, or in other cases a cold smirk. This unfortunate reality, which includes epidemic vitamin and mineral deficiencies of young brains, often leads to confusion, mental aberrations and a purposeless life with no hope and no moral compass in these young men. Even Proverbs (4:23) says "Keep thy heart with all diligence; for out of it are the issues of life." Time magazine (April, 2018) and recently many others have even reported on widespread depression on college campuses, showing that record numbers of college students are seeking treatment for depression, loneliness and anxiety, and many of these schools are simply not yet equipped to handle this. Again, what is going on? Again, same problem.

Much of mainstream American medicine as well as many educational and media elites ignore or disagree with what I've just described and promoted. People need to wake up! We have an unprecedented and tragic problem. Many decades ago in the my mountainous and scenic home state of West Virginia, where I grew up nearly always outside, and often in the woods, most boys couldn't wait to get a BB gun, and later a .22 rifle. Once of age you could purchase a more serious gun without paperwork, registration, or background checks either through the mail, or from the local hunting and fishing store. It was no big deal. That's the way my grandfather and favorite uncle did it in West Virginia (I still have some of their cool hunting license badges from the 1930's and 40's), and myself as well as a young man. There were no mass shootings. Guns have always been around; they are just a small part of the problem.

A troubled 19-year old identified as John Earnest recently stormed into a synagogue (April 27, 2019) near San Diego and killed a much-loved 60-year-old woman and wounded three others, including the rabbi. Sadly, a few days later and just forty minutes away from my home in nearby Charlotte, N. C., another troubled young man opened up with a pistol at the university, killing two other young men and wounding several others. One of the fatalities was an intrepid young man named Riley Howell who charged and tackled the troubled shooter, saving other lives, but he himself took three bullets in the close quarters encounter. Two days later on TV I witnessed the gripping scene of his grief-stricken father slowly saying this: "My son had the right to *go to school*, to *learn at school*, and to *be safe at school*." It was heart-breaking to watch.

I heard on the news the other day an announcement about new mass shooting "preparation" seminars, and the exploding sales of bullet-proof backpacks. What a different country we live in today, compared to 25 years ago. Schools have changed quite a bit from my day when we had discipline, occasional prayer, loved the flag, and the biggest problem was chewing gum and spit wads. It's hard to even keep pace with the stream of shootings, writing this book. I'm returning to this paragraph to add a mention of the worse-than-tragic mass shooting that just occurred in Virginia Beach, where a male city employee of the Public Utilities Department opened lethal indiscriminate fire immediately after entering a large building in the city's Municipal Center, killing eleven civil servants and one contractor who was there for a permit. Please consider rereading the previous few pages and try to imagine the carnage and destroyed families. If nothing changes, do you really think the mass shootings will stop?

As I near the completion of this book, I am sadly returning to this chapter, yet again, as two other lost young men apparently in their early 20's have committed mass murder yesterday within fourteen hours of each other, in Dayton, Ohio and El Paso, Texas (Aug. 3rd-4th, 2019). As I earlier said, it is hard to keep up with. It appears right now as the information comes in that in Dayton's historic Oregon restaurant/bar district the casualties total 10-12 dead and 26 injured or wounded, while the totals from the El Paso Walmart massacre are 20 killed and about another 26 injured. I'm tired of using the word tragic. This is **catastrophic** . . . but these days, not unexpected. When I thought I had completed this chapter in the book you're now reading four months ago, I closed with the question, "If we change nothing, do you really think the horrific shootings will stop?" My answer then: "Hardly a chance." Now, it is obvious the proper answer is "**absolutely, unequivocally NO**, the mass shootings **will NOT stop**," if nothing changes. Again, guns are but part of the problem. Chicago is the murder capital of the country despite tough gun laws. Obviously most bad guys couldn't care less about such gun laws, unless the result is de-armed citizens; then they approve. We do not just have a new, serious societal problem here. We are in the midst of an unprecedented historic crisis.

37

WHITE LAB COAT INTIMIDATION, PEER PRESSURE, AND EGREGIOUS EISEGESIS

THEISTIC EVOLUTION IS the belief that God used evolution as the process to bring forth His creation. Followers of historical, biblical, orthodox Christianity and many conservative Christians today reject the theistic evolution idea because, among other things, it attacks and refutes the clear description in the book of Genesis that Adam, the first man, was made in the image of God, from the dust of the earth, and then Eve from his side (Hebrew, *tsela;* Genesis 2:21*)*. The historicity of this amazing event is confirmed in the New Testament in First Corinthians 11:8, and First Timothy 2:13. Ever since Adam (over six thousand years ago), men and women have obviously been born of women. In the beginning, however, the first woman was made from man (Genesis 2:21-24). There is no scientific, biological, conceivable way in which evolutionary processes could first form men, and then women. Following theistic evolution thought, both male and female humans miraculously evolved *simultaneously* from an imaginary population of hominids, but the above verses, and real science refute that notion (see also *The Male-Female Evolution Enigma*, Part II, Ch. 24). The first man and woman were uniquely created by God in His image, not evolved by random chance from ape-like progenitors.

Theistic evolution undermines many other important biblical principles, including the all-important concept of death entering God's "exceedingly good" created world through the rebellion and actions of one man, the *first* Adam. According to this strange attempt to mesh long ages (deep time) evolution with the Bible, Adam and Eve supposedly evolved from ape-like creatures, even though there is **no known scientific evidence supporting that view**. As a matter of fact, the unsupported ape-to-man evolution idea is filled with not only a plethora of insurmountable and even unimaginable biological, physiological, and morphological challenges, but with no needed transitional forms *ever found* in the fossil record. What we have is a century of misinterpretation, deception, dishonesty, and *even outright fraud* (Java Man, Piltdown Man, Nebraska Man, Peking Man and even to some degree little chimpanzee-like *Lucy*) by evolution proponents as they desperately searched for the supposed ape-to-man missing links, and the hoped-for fame, status, and grant money to follow.

The theistic evolution idea, popularized in the last two hundred years but particularly catalyzed after the publication of Darwin's famous 1859 *Origins* book, assails the character of God by essentially blaming Him for untold hundreds of millions of years of struggle, suffering, bloodshed, predation, disease and death . . . almost as if He was experimenting and couldn't make His mind up—over millions of years—with this horrific, inefficient, sanguinary, trial-and-error process. Furthermore, why did Jesus have to die, and how can His death pay the penalty for sin, if death has *always* been around? It negates the message of the atonement. The horrific theistic evolution scenario is also *not* one most people would categorize as "very good." This is in strong opposition to the clear words in Genesis where it distinctly describes the sequence and focus of the *six* 24-hour creation days, six times stating in Genesis 1 God's creation work to be "good," and then when finished—including the creation of "the host of heaven" (Genesis 2:1) and man on day six—God subsequently pronouncing it all "*very* (or *exceedingly*) *good*" (Genesis 1:31). The just described wasteful and exceptionally cruel billion-year theistic evolutionary proposal could hardly be described in such terms.

On top of that, if theistic evolution is indeed the method God used, it should totally negate and discredit the God of the Bible as a God of order and efficiency, no mistakes, power and intelligence, as well as love, grace and truth (with the obvious inability to clearly say what He means in Scripture). It is important to understand there is *not a hint* of long ages (millions or billions of years) anywhere in the Bible. Additionally, Christians who accept the evolutionary idea, which goes against the **no exceptions to** 1st and 2nd Laws of Thermodynamics, Information Science, and the Law of Biogenesis—just to name a few— must accept that evolution is *still* occurring today, **including in human beings**. Interestingly, even famous Harvard evolutionary paleontologist Stephen Jay Gould surprisingly stated: 'We're not just evolving slowly. For all practical purposes we're not evolving. There's no reason to think we're going to get bigger brains or smaller toes or whatever—we are what we are.'[187] But theistic evolutionists, who include many compromising Christian pastors and seminary

---

[187] Stephen Jay Gould, (former Professor of Geology and Paleontology, Harvard University), in a speech in October, 1983; as reported in 'John Lofton's Journal', *The Washington Times,* Feb. 8, 1984

professors, and leaders of well-funded organizations like the BioLogos Foundation, simply *cannot* rightly say that God formerly used evolution and now doesn't. That would negate evolutionary belief, for along with the often ignored reality that evolution has **never been observed** in the present, *nor* in the past via the fossil record— such a claim that God does *not now* use evolution would absolutely invalidate their evolution idea.

Compromising Christian leaders have always been around, even going back to the first century, but their numbers increased exponentially after the publication of Charles Darwin's *Origins* book in 1859. Reinterpretations of Genesis—including the creation account, Noah's Flood, and the Babel dispersion—were the focal point of such compromise and eisegesis, as these men were powerfully influenced by the now known-to-be **false** assumption that Lyell and Darwin and others had proven evolution and a millions-of-years timeframe (early *white lab coat intimidation*). This even included famous Presbyterian theologian Charles Hodge (1779-1878) of Princeton Seminary, who succumbed early on to the *gap* theory, but then switched later to the *day-age* theory. The dominant thought process then, though not validated by any real scientific evidence: 'We just have to come up with something plausible to squeeze in millions of years; the geologists have apparently proved them . . . I guess.' I still have a worn, annotated *Schofield Reference Bible* in my collection. C. I. Schofield's very popular century-old *Reference Bible*—used by millions—shows endorsement of the *gap theory* in his notes on Genesis 1:2, as he *fluidized* the Genesis historical narrative.

Fluidized is probably not in your dictionary, but it's my word for what happens to the words and meaning of scripture, especially the obvious historical narrative of Genesis, when the author's clear, intended meaning is gouged by critics or compromising Christians. That allows denial of what Scripture actually says and means, reinterpreting the scriptures to suit one's personal own views and opinions. This subsequently leads to similar destruction, melting, or metamorphosis of other Bible verses to mean whatever one wants them to mean (it happens all the time in this postmodern age). I could list many more well-known clergy victims of peer pressure and white lab coat intimidation, but you get my point. This compromise, disrespect of God's Word as written, and reinterpretation of the foundational Genesis creation account—*egregious eisegesis*—was (and still is) guided by the belief and *assumption* that geologists and other scientists have proven millions of years, also referred to as *deep time*. In other words, these Christian leaders and pastors were (and still are) influential victims of *white lab coat intimidation*. Very little could be further from the truth than the validation of deep time (see most of Part II, and Part III- *The Problem with Assumptions, Ch. 29*). Mountains of scientific evidence—some discussed in this book—not only agree with the scriptures, but **scream *young earth***.

> Let's define our terms:
> eisegesis: the process of interpreting text in such a manner that allows or favors *one's own* presuppositions, biases, worldview or agenda; applied to often legitimize or 'prove' a pre-held position or opinion by the reader, writer, historian, scientist, or critic
> exegesis: critical and honest interpretation of the text— especially applied to Scripture—where the scholar or reader attempts to determine the *intent and true meaning of the original author*

BioLogos is a foundation that strongly promotes and encourages Christian churches, professors, authors, and other believers to deny the foundational Genesis creation historical narrative, and instead adopt and teach evolution and millions of years. They have garnered millions of dollars from organizations with deep pockets, including the Templeton Foundation, which also funds other groups to get Christians to embrace evolution. These other programs with their accompanying significant stipends or grant money, often awarded to Protestant or Catholic seminaries, support new programs ostensibly to help them better integrate science into basic doctrinal and theological teaching. *Translation*: to destroy, morph, or *fluidize* the Genesis historical narrative and promote theistic evolution. In 2014, their then (and current) president Deb Haarsma claimed that churches that supported evolution would be more effective witnesses in a science-oriented culture, and their college students would be better prepared to compromise, avoid conflict and avoid a potential faith-shaking crisis when challenged by evolution-based science professors. Just stunning—this attitude is one of the main reasons we're "in the mess we're in" (I wonder if she has ever thought deeply about I Peter 3:15, or Revelation 22:18,19)? Christians that wimp out, or promote a false *apologia,* irk me. The same applies to nauseating "lukewarmness" (Revelation 3:16).

This obvious misguided policy by Ms. Haarsma, which I gleaned from a Nov. 29, 2014 WORLD Magazine commentary,[188] was followed by another misleading, unsupported statement from the BioLogos website: "Genetic evidence shows that humans descended from a group of several thousand

---

[188] Daniel James Devine, *Interpretive Dance*, WORLD Magazine, Nov. 29, 2014, p. 35

individuals who lived about 150,000 years ago." This is a flat-out unproven claim as well. It is ironically demonstrated in this book that evolution acceptance and teaching, and egregious eisegesis—*fluidizing*[189]of the Genesis text by pastors and Christian leaders—is one of the main reasons for **the loss of biblical authority** and the deep cultural chasm across America today. Sharp young people can recognize **fakes and compromise**. In dealing with the issue, and considering the opposing viewpoints, and the fluidizing of the Genesis text, here's their common thought process and question: "If I can't trust the foundational claims and history in the Bible's beginning foundational book of Genesis, then what part of the Bible can I trust?" Or, "Why should I trust *any* of it?" Such are legitimate questions, indeed. I had them as an indoctrinated college boy. Recall my earlier reference to Jesus' response to Nicodemus (John 3:12): "If I have told you earthly things, and ye believe not, how shall ye believe, if I tell you of heavenly things?"

As this book nears completion, I hope the informed reader now realizes that the various compromise positions on the Genesis creation historical narrative (theistic evolution, the gap theory, the day-age theory, progressive creation, the framework hypothesis, and a few other oddballs), do have a common denominator. They all attempt to squeeze millions and billions of years of earth history (*deep time*) somewhere into Genesis 1. You young adults out there, be smarter than I was, and always stay on propaganda alert. Do you still retain that lingering question (like I did as a young fellow) that goes something like this?

> But, my goodness, with a lot of the professors and scientists apparently believing and teaching it, and these organizations and foundations supporting it, and still promoted in most of the museums, textbooks, and movies . . . doesn't evolution have to be true?! (And doggone it, I don't want to be mocked, or made fun of . . .)

Consider the following quote by intellectually honest British zoologist and marine mammal expert L. Harrison Matthews, who was given the honor,

---

[189] "fluidizing" the Genesis text, and/or other key passages of the Bible, unjustifiably changing obvious historical narrative into some other literary genre or classification- like allegory, poetry, myth, or whatever- allows the reader to give the particular text *any* meaning he or she so desires . . . what I think of as not only egregious, but *extreme eisegesis*

so to speak, to write the introduction to a new, much anticipated 1971 version of Charles Darwin's *Origin of Species*:

> The fact of evolution is the backbone of biology, and biology is thus in the peculiar position of being a science founded on an unproved theory—is it then a science or a faith? Belief in the theory of evolution is thus exactly parallel to belief in special creation—both are concepts which believers know to be true but neither, up to the present, has been capable of proof.[190]

I do so appreciate when scientists or evolutionists speak with truth, irrespective of their personal worldview preferences. Beware of theistic evolution—perhaps the worst of the Genesis compromises—and do not be misled. Consider the following two logical reflections by internationally recognized and respected scholars with doctorate degrees, albeit with opposing worldviews, but who make similar points:

> Evolution is the cruelest, most wasteful and most irrational method of "creation" that could ever be imagined, not even to mention the fact that it is scientifically untenable. The postulated suffering and death of multiplied billions of animals in the course of evolutionary "progress" from amoeba to man is a libel against the character of the Creator—who must certainly have been capable of creating each organism complete, with its own perfectly designed structure for its own unique function, *right from the start*.[191]

That commentary is from brilliant **Bible scholar and teacher**, historian, author, engineer, and scientist Dr. Henry Morris (1918-2006). Note the similarity of reasoning from *atheist* and Nobel Prize-winning French biochemist Jacques Monod (1910-1976), commenting on the challenge of meshing natural selection and evolution with Jesus and Christianity:

---

[190] L. Harrison Matthews, FRS (Fellow of the Royal Society) , Introduction to Darwin's *The Origins of Species*, J. M. Dent & Sons Ltd, London, 1971, p. xi

[191] Ibid, Morris, *The Long . . .* p. 58

And why would God have to have chosen this extremely complex and difficult mechanism? . . . Why not create man right away, as of course classical religions believed? . . . [Natural selection] is the blindest, and most cruel way of evolving new species . . . The struggle for life and the elimination of the weakest is a horrible process, against which our whole modern ethics revolts . . . **I am surprised that a Christian would defend the idea that this is the process which God more or less set up in order to have evolution.**[192]

Although downplayed and disrespected in our post-modern world, sound *exe*gesis is important and necessary. As a truth detector, though, be on a constant lookout for *ei*segesis—those personal opinions and interpretations of critics and pundits—in opposition to the author's intended meaning, especially related to study of biblical passages. It would be interesting indeed to hear Francis Collins and Deb Haarsma of BioLogos comment on—and attempt to refute—the logic in these two scholars' commentaries.

---

[192] Jacques Monod, *The Secret of Life*, interview with Laurie John, Australian Broadcasting Co., June 10, 1976

PART IV

BRIEF MENTIONS

38

A LITTLE HEALTH TRUTH

I AM WORKING ON a small and (what I consider) important book called *Health and Wellness Truth . . . and Misinformation*, but I need to at least touch on the subject here, as it is so important. Mistreatment, over-treatment, deception with Big Medicine/Big Pharma, widespread chronic disease, excessive and sometimes dangerous scans, questionable testing and vaccinations as well as high prices are all part of American medical treatment today. The huge overuse of dangerous drugs—especially anti-biotics—is part of this as well.

First things first: America consumes junk, and is unhealthy and outa shape. As 2019 closes, 42% of Americans are obese, and it was recently announced that fewer and fewer candidates qualify for military service because of it. This is downright sad and embarrassing. Diabetes is rampant, and over 90% of the time is caused by poor diet and lifestyle choices—period. We are the unhealthiest of all western industrialized nations as many Americans feast on refined sugar, carbohydrates and depleted wheat, shun pure water drinking while consuming tons of unhealthy soda and carbonated beverages, and live off *fast* and *processed* foods (for more info, see Ch. 36). Studies show the average American consumes 152 lbs. of sugar per year (25 lb. of that being candy), compared to 1 lb. per year in 1900. No wonder we have sick people everywhere. You are what you eat. Non-health and disease abound. The popular but terrible American diet combined with lack of exercise has led to a plethora of problems besides diabetes: heart problems, gut/digestion/reflux problems, and osteopo-rosis/arthritis/lupus. Poor nutrition (and drugs) can be involved in various cancers, skin disorders, ADD/anxiety/depression and has contributed to Alzheimer's and dementia. Toddlers and kids are sick all the time with colds, sore throat, flu, sinus, allergy problems and much more. The cost, both in quality of life and financial, is staggering.

Limited interviews and evidence indicate that most of America's dis-turbed young men involved in the horrific mass shootings the last two decades generally had a terrible fast/processed food, sugar-laden nutri-tionally-deficient diet. Such a chemicalized, non-food diet combined with daily carbonated drinks starves the brain for nourishment and nutrition, particularly essential fatty acids and the all-important B-vitamin complex. If the American people would switch to a healthy whole-foods diet, or something akin to the Mediterranean diet, avoid fast and processed non-foods, drink primarily pure water, shun refined sugar, refrain from smoking and exercise or walk regularly, a good number of American medical businesses would have to close their doors.

We are hooked on prescription drugs, all of which have potentially bad side effects (they're not really side effects; they're *effects*). And when you're taking three or more, as many Americans do, there is not a pharmacologist or biochemist around that can predict the particular bad reactions (or side effects) on you; everybody is different. The pharmaceutical industry is so corrupt it would blow most people's minds. I keep track of the big-big fines levied against Big Pharma. It looks like the latest one—some $12 billion . . . yes, around TWELVE BILLION Dollars—laid on Purdue Pharma, tops them all (relative to opioid indiscretions). Many over the counter drugs are not safe either, especially if you take them regularly. Everyone's goal should be to take zero prescription drugs, or as absolutely few as possible.

Dr. William C. Douglass, an American physician who sometimes bucked the establishment, said this: "While our bodies may have come with an expiration date, the truth doesn't." There are simply certain basics, or truths, that apply to good health and wellness. There are no shortcuts, no magic pill, and it truly is not rocket science. Or, as the great grid coach Vince Lombardi used to say, "There are no new fundamentals." I believe the negative trends with Big Medicine/Big Pharma and our health parameters are even indirectly related to the larger culture war. This book doesn't allow an in-depth discussion on details and specific health conditions, treatments, antibiotic and vaccination abuse, over-used scans and tests, but my follow-up *Wellness* book does.

Remember that **you** are in charge of your health-not some doctor, the FDA, your insurance company, the latest-greatest magic pill, some entitlement program, or even Google. Wellness-wise, you want to do the best you can with the cards you've been dealt, with no dementia, Alzheimer's, or years languishing in a nursing home. Understand that many physical problems, with irreversible *structural* change, simply follow *functional* change that has been going on for many months or years.[193] That is often related to poor diet and lifestyle choices. Natural healthcare teacher, writer, chiropractor and applied kinesiologist Dr. Robert Blaich says it best: "Creating

---

[193] William Campbell Douglass II, MD, *Into the Light*, Rhino Publishing, Panama, 2003, p. 3

health has to take precedence over treating disease . . . it is that simple."
Absolutely! The goal: to live a long, productive and active life with vigor,
and when your time comes, your candle just flickers a time or two (no
languishing in nursing or dementia homes) and then goes out. More on
this to come.

39

WHEN THE BIBLE *TOUCHES ON* SCIENCE

THE BIBLE IS not a science book per say, but when it *touches on* science, it is profoundly accurate and even prophetic. As one who has always been keenly interested in the workings of science and study of the natural world, I continue to be impressed with the amazing anticipations of modern science revealed in the Scriptures. Space constraints allow just a handful of examples here (from among many), but you can have increased confidence that what we *read* in God's Word always agrees with what we *see* in God's world.

> Psalms 8: 8; "The fowl of the air, the fish of the sea, and whatsoever passeth through the paths of the seas." Virginian Matthew Fontaine Maury (1806-1873) served land duty with the Confederate Navy after 36 years of service in the United States Navy, and was experienced in cartography, astronomy, geology, meteorology and oceanography. During convalescence while recovering from an injury, a visiting family member offered comfort to Maury, a strong Christian, by reading from the book of Psalms. When he heard the phrase "the paths of the seas" (Psalms 8:8), he immediately perked up, and his thoughts repeatedly returned to that verse. Mr. Maury, a strong Christian, developed conviction that if God said there are *paths in the seas*, then they must be there, and he was going to dedicate himself after his recovery to find them. For years Maury collected thousands of ship logs and information from sea captains from around the world and studied their notes, weather, ocean current and travel information and conducted many interviews, confident that what God had said was true. He came to an even stronger understanding beyond that obtained from his previous decades of naval service, that

the oceans had regular circulating water currents interacting with wind, land, and salinity input and that this knowledge could be extremely beneficial to maritime businesses and sea captains, promoting safer and more efficient ocean travel, and other benefits. His eventual magnum opus on oceanography, The *Physical Geography of the Sea* (1855), from this Scripture-inspired goal, is still considered the classic work on the subject. The human and monetary benefits to shipping, sea travel and oceanography from Maury's life work are inestimable. The wonderful book titled *Matthew Fontaine Maury: Pathfinder of the Seas*, by C. L. Lewis, was issued by the United States Naval Institute in 1927. The Bible *touches on* science in Psalm 8:8, Matthew Fontaine Maury *believed*, and the entire world has benefitted for almost 150 years.

Deuteronomy 23: 12-13; "Thou shall have a place also without the camp, whither thou shall go forth abroad: And thou shall have a paddle upon thy weapon; and it shall be, when thou ease thyself abroad, thou shall dig therewith, and shall turn back and cover that which cometh from thee." This scripture is directed toward the Israelites in the wilderness after the delivery from Egypt. There were perhaps several million of them, and human waste disposal was important. God cares about our health, even to the point of giving instructions on waste disposal (human excrement), to be performed outside of camp. Such strict practices would drastically cut back on dysentery, cholera, and other diseases. This was written over three millennia ago, but historically, more wartime fatalities occurred from infection, poor sanitation, and neglecting proper management of human waste than directly from battlefield bullets and bombs up until WW1. This was definitely the case in the American Civil War (well over 600,000 casualties).

Genesis 1: 11, 12, 21, 24, 25 (and another half-dozen places); "after his kind," or "after their kind"; this much-repeated important phrase refers to the restricted ability for all of the different created animal and plant types to reproduce and/or proliferate only within refined, specified limits (now known to be determined and programmed by the genetic code (DNA), always producing offspring like the parent organisms. Biology now confirms that creatures reproduce *only within their own kind*, with minor variations (like coyotes and jackals, or leopards and jaguars, or within Darwin's Galapagos finches), but never changing into a different type—or kind—of organism. After centuries and millennia, dogs are still dogs, cats are still cats, and finches are still finches—even on the Galapagos Islands.

Acts 17: 26-"And hath made *of one blood* all nations of men for to dwell on all the face of the earth." Modern genetics and biochemistry has confirmed there is *only one* type, or *race*, of human beings. There are different skin colors, but that is due merely to different amounts of the same pigment—melanin—in the skin. The term race is *never* used in Scripture; the Bible simply talks about tribes, people groups and kingdoms. The unfortunate and odious word *race*, as commonly used the last two centuries, **is simply an offensive *evolutionary construct*** denoting a group (even a particular *human* group) or population evolving into a new species, often with implications of inferiority or superiority. How better off the world would be, if we all truly realized, as God's *image-bearers*, that we're simply *of one blood*, and the **only race is the human race** (notwithstanding evolution claims, and past horrific *Social Darwinism* policies). Or, as African-American community organizer Ann Atwater (actress Taraji P. Henson) said to Ku Klux Klan chapter president C.P. Ellis (actor Sam Rockwell), in the fine 2019 film, *The Best of Enemies* (on the desegregation of the 1971 Durham public school system), "Same God made you, made me."

Genesis 1: 2, 6, 7, 9-21; these verses have to do with the earth being the *water* planet, with water the foundational starting point and compound, when "the earth was without form, and void" (Genesis 1: 2). Dr. Russell Humphreys, a young-earth creationist Christian and with a Ph.D. in physics, was fascinated by these beginning verses and went to work years ago to produce a model to hopefully predict the magnetic fields of the planets. He went from the biblical starting point, using physics and mathematics, that the planets were "**made out of water**," and calculated their initial magnetic field at the point of creation. Then Dr. Humphreys further applied his model and calculated the expected magnetic field strength of the planets, *if* this creation and solar system were only 6,000 or so years old, which is what the genealogy timeline of the Bible (and emerging scientific and genetic evidence) indicate. Using this model, and beginning with a *touch on science* in Genesis 1:2 and some following verses, Dr. Humphreys **amazingly predicted the magnetic field strength of the planets Uranus and Neptune** before the Voyager spacecraft measured them. He **also predicted that Mercury would still have a magnetic field** when all other models (which assumed eons of time) predicted that it should be long gone. Likewise, still crunching the numbers following his model, he predicted that **Mars would have remnant (permanent) magnetism**, which was also unexpected by the secular scientists. In **every single prediction, based on biblical starting assumptions of a recent creation—and from water—some 6,000 years ago**, Dr. Humphrey's model **was stunningly accurate**. If the planets were really billions of years old, their magnetic fields should be extremely weak, but they are not. These magnetic fields confirm formation, or *creation*, of the solar system and planets **about 6,000 years ago**.[194] Regrettably, it is not taught in public schools or college science classes because of bias and censorship; it refutes the entrenched, billions of years, secular evolutionary paradigm.

Job 38:16-"Hast thou entered into the springs of the sea?" This question was posed to Job by God some 3,000 years. It is only relatively recently that scientists have discovered these "springs" on the sea floor.

Proverbs 17: 22- "A merry heart doeth good like a medicine; but a broken spirit drieth the bones." What did King Solomon

---

[194] Much of this from *Inspired Evidence*, Julie Von Vett and Bruce Malone, Search for the Truth Ministries, Midland, MI, 2011, Nov. 10th page; taken partially from *Taking Back Astronomy*, by Jason Lisle, 2006, p. 63-64, and *What Does the Bible Say about Astronomy,* Jason Lisle, 2005, p. 11-12. Additionally, Dr. Russell Humphrey's *Starlight and Time*, an outstanding easy-to-read short book concerning the distant starlight issue and time dilation, related to recent creation, is highly recommended.

imply in this statement some 3,000 years ago? We now know that *depression* is a risk factor for osteoporosis; it somehow stimulates inflammatory chemicals to leach away some of the bone components (*author's warning: smoking, alcohol, and regular carbonated beverages consumption can do likewise). It seems King Solomon was right, nearly 3,000 years ago . . . as was, more recently, singer-songwriter Bobby McFerrin, through his song, "Don't Worry, Be Happy."

<u>1st Corinthians</u> 15: 41-"There is one glory of the sun, and another glory of the moon, and another gory of the stars: for one star differeth from another star in glory." As man gazes upward at night, all stars look alike, including in the Palestine area 2,000 years ago. Even looking through a telescope, they still just appear as points of light, some a bit brighter than others. But here we have a profound modern confirmation of biblical revelation, as it is now known that all are different, with no two apparently alike. This is revealed through modern high-tech analysis of light spectra, with the discovery of each point of light's particular distinctive ratio of brightness to temperature. They each reveal their own spectroscopic fingerprint, so to speak, and are indeed different . . . in *glory*. There is no other religious book remotely like the Bible. When it touches on science, it is amazingly prophetic and accurate.

<u>Job</u> 38: 29-This is another rhetorical question directed at Job by God: "Out of whose womb came the ice? And the hoary frost of heaven, who hath gendered it?" At this point in ancient history (Job's era), the Ice Age would have been in full swing (see the section, *The Ice Age: You Need Cold and Hot*). Although Job apparently did not live in a frigid, icy area, he almost certainly would have heard of the huge ice sheets and snow to the north from travelers. Those of you that have had the unique experience of walking on glaciers in the Rocky Mountains know that Scripture's description of ice slowly coming forth, as if from a womb, is an appropriate and accurate picture of a real glacier.

<u>Leviticus</u> 15: 11-(NIV) "Anyone a man with a discharge touches without rinsing his hands with water must wash his clothes and bathe with water, and he will be unclean till evening." The entire fifteenth chapter of Leviticus is on hygiene, but unfortunately these wise guidelines were often ignored in recent centuries. In the 1800s, Hungarian physician Philipp Semmelweis discovered that infection, particularly puerperal fever, could be drastically diminished by the use of thorough hand washing in obstetrical medical clinics. He has been described as the "Saviour of mothers," but unbelievably many doctors rebelled against the nuisance of repeatedly washing their hands, and Semmelweiss was fired. Unsurprisingly, the death rate increased drastically. Doctors not only ignored Biblical hygiene commands, but seemed to take pride in their bloody hands and lab coats. This is even somewhat apparent with the bloody hands in one of the greatest of all American paintings, *The Clinic of Dr. Gross* (or simply, *The Gross Clinic*) by Thomas Eakins, in 1875. Finally about this time, at least a decade after the American Civil War, the medical industry reluctantly acknowledged that hand-washing and antiseptic procedures did prevent disease, but untold thousands had already perished needlessly, because of neglect of biblical guidelines. Many such guidelines and hygiene laws (Scripture *touching on* health and medical science) are found throughout the books of Leviticus and Numbers, given over 3,000 years ago.

<u>Jeremiah</u> 33: 22- "As the host of heaven cannot be numbered, neither the sand of the sea measured," refers to the uncountable stars and heavenly bodies in the universe. During Jeremiah's time only some 3,000 could be seen with the naked eye. The ancients had no knowledge that there were billions of *galaxies—each containing many billions of stars—* virtually equivalent to the sand grains of the ocean and seashores. This was first discovered 22 centuries later by Galileo and his telescope, then confirmed more strongly by our modern equipment, but this Biblical/scientific truth was revealed by God over 2,500 years before.

<u>Hebrews</u> 11:3-"Through faith we understand that the worlds were framed by the word of God, so that things which are seen

were not made of things which do appear." Today scientists not only know that all matter is made up of unfathomably tiny particles referred to as atoms, but also that the individual atoms can have as many as eighteen subatomic particles. Of course, these super-nano particles cannot be seen ("things which do [not] appear"), but the physicists infer they exist by the effects they produce. Only the Creator-God could have known, twothousand years ago (when Hebrews was written), that the physical world around us—the visible, real matter—was composed of invisible subatomic particles. Going further, Hebrews 11:1 says "Now faith is the substance of things hoped for, the evidence of things not seen." We have faith in an awful lot of stuff that falls under "things *not seen*," like magnetism, gravity, electricity, atoms, or the inner workings of the jet engine in the aircraft carrying me from Charlotte to Austin, just to name a few. Richard and Tina Kleiss make a great closing point:

> Similarly, we cannot see God with our eyes. We can see his overwhelming power, design, complexity, and beauty displayed throughout the universe, however, from the smallest bosons and quarks (subatomic particles) to the largest galaxies. God has given us abundance evidence to validate faith in Him (Romans 1:20).[195]

Psalms 139:13-16-Here David discusses the miraculous formation of a human infant, claiming "thou hast *covered* me in my mother's womb," where covered can be beautifully translated as *shielded*, referring to the structure and design of the amazing nourishment and protection system of the tiny, growing child. Then he adds he is "fearfully and wonderfully made," where wonderfully essentially means *differently*. That is, all baby humans are generally the same, but different-**special and unique**-in the details (kind of like snowflakes and buck antlers, only more so). Finally in verses 15 and 16, David states that he was "made in secret, and *curiously wrought* . . . and fashioned," which means *embroidered*, which is a

---

[195] Richard and Tina Kleiss, *A Closer Look at Prophecy*, Search for the Truth Ministries, Midland, MI, 2019, Apr. 7

phenomenal and appropriate description of the now well-known, controlling DNA double-helix molecular structure, along with its accompanying RNA and protein enzyme workers and assistants, and its mind-boggling protein synthesis and life-building mechanism. Here we are talking about virtually incomprehensible, all-world, gold medal-winning *embroidering* (way beyond my talented great aunts from the West Virginia hills). King David had his faults, for sure, but what an amazing insight from 3,000 years ago. In closing, I consider it significant that this same guy "after God's own heart" says in verse 16 from this passage that "Thine eyes did see my substance, yet being unperfect; and in thy book all my members were written, which in continuance were fashioned, when as yet there was none of them." One important point here concerns the words "substance, yet being unperfect," and the use of the word *un*perfect—not *im*perfect. Of course the Old Testament was originally written in Hebrew, and this phrase is translated as just one word in the Hebrew meaning *embryo*. The first part of verse 16 indicates that God is watching over each miraculously (and slightly *different*) tiny embryonic human being from the get-go, **from the very moment of conception**, already with an eternal soul . . . just *un*perfect until birth. What an amazing "touch on science"—and more—from King David.

40

NICODEMUS: WAS HE SMARTER THAN A 5TH GRADER?

IN CHAPTER 3 of The Gospel According to John, Nicodemus, a member of the Sanhedrin, which was the Jewish governing council, came to see Jesus at night. It was apparently not because he was afraid, or didn't want anyone to see him associating with or endorsing Jesus (as later events would prove); it was because . . . well, he was *feelin' it*—he knew this Jesus guy was something *special*. Crowds of folks were all around during the day, and Nicodemus, a godly and learned man, who had earlier been prepped—or catalyzed—by John the Baptist's messages, wanted some one-on-one time.

After respectfully addressing Jesus as Rabbi, the scholarly Nicodemus starts right in firing questions. Jesus answers him simply by talking about being **born again**, and refers to regeneration by the Holy Spirit (what I think of as *getting a new heart*), and Nicodemus is puzzled; he just doesn't quite get it. In fact, in John 3:10 Jesus, in a polite, seemingly mild rebuke, asks him "Art thou a master of Israel, and knowest not these things?" I was almost chuckling in thinking about Jesus' response recently; my youngest son in former days probably would have asked, "Does that dude have cobwebs in his brain?!" Then I recalled the show "Are You Smarter Than a 5th Grader?"

The word *master* (KJV) used here could literally be translated "*the teacher*"; one could make the argument that Nicodemus was probably *the* most well-respected Old Testament scholar among the Jewish religious leaders. But he was struggling to comprehend the simple straight-forward message of Jesus (as many educated, scholarly types do), who was essentially saying that even righteous, learned Nicodemus—perhaps the greatest teacher in all of Israel—was born with a sin nature, and must be born again (get that *new heart*) to enter the kingdom of God. Of particular interest to me a few verses later (almost as if He is indeed addressing a 5th grader), in John 3:12, Jesus significantly asks "If I have told you earthly things, and ye believe not, how shall ye believe, if I tell you of heavenly things?" This is so relevant to the culture war America is immersed in today, and is a big part of the problem. The undeniable reality:

> If men will not believe the "earthly things" of the Bible (Creation, Flood, dispersion, etc.), which human records and research can verify, then why should they believe the "heavenly" things it speaks about (salvation, heaven, eternal life, etc.) which must be accepted *strictly on faith*. Many modern evangelicals have become involved in this inconsistency, rejecting the earthly things while still credulously accepting the heavenly (spiritual) things.[196]

Indeed some have, mostly due to white lab coat intimidation, German "higher criticism" (see Part III), and stifling academic enforcement of *only* the Darwinian *deep time* worldview allowed. Dr. Henry Morris again, from over thirty years ago:

> This is the underlying reason why most young people have rejected not only the scientific authority of the Bible but also its religious and moral authority, and along with that the authority of the church, the school, the home, and everything. Each person has become, in effect, his own god, with his own self-determined standards of truth and morality.[197]

Indeed, everything is *relative* now. Don't fall for that, and be smarter than a 5th grader. And by the way, Nicodemus came around in a big way with fellow Christ-follower Joseph of Arimathea, a wealthy man and also a member of the Sanhedrin, who even purchased rocky ground in Jerusalem for a tomb for Jesus. It makes no sense that he did this for his *own* future burial; he didn't even live in Jerusalem. Then there is the question of why *in the rock*, right near Golgotha, close to the crucifixion site? Indeed the description of this in the Gospels strongly hints that Joseph of Arimathea

---

[196] Dr. Henry Morris, annotations, *The Defender's Study Bible*, World Bible Publishers, Inc., 1995; John 3:12 note, p. 1137

[197] Henry M. Morris and Martin E. Clark, *The Bible Has the Answer*, Master Books, Green Forest, AR, 2003 (Fifteenth printing), p. 7

(possibly with Nicodemus' assistance) bought and prepared this sight ahead of time to receive the body of Jesus. It's even possible he was in the tomb himself, watching and listening during the crucifixion, as when Jesus died Joseph "went in boldly unto Pilate, and craved the body of Jesus." Joseph moved fast; Pilate was stunned that he might already even be dead. These two dudes were at some risk for requesting Jesus' body, but likely couldn't have cared less. It will be great to meet them in heaven.

The Bible touches on those who rely totally on logical analysis, reason, their graduate degree, or intellectual acumen to consider the claims of Jesus and the Gospel message, as Nicodemus may have initially tried to do. I'm embarrassed to admit it, but it was the same for me many years ago, when still enamored by the rarified air of higher academia and with a mild infection of the *status disease*. These truths are spiritually discerned. First you've *got to get your heart right*. As Jesus told the Hebrew scholar that night, you've got to be "born again." Is Jesus saying you have to give your heart to Him, or he will change your heart? Oh no, it is much better than that . . . if you *receive* Him, He will actually give you a *new* heart.

41

TOLEDOTH . . . WHO WROTE THE EARLY CHAPTERS OF GENESIS?

IN THE GLOSSARY at the end of this book, I have included a definition and discussion about the Hebrew word for "generations," toledoth, which in the Septuagint (the Koine Greek Old Testament) is translated *genesis*. Consider taking a look at that, as well as Chapter 33 (Part III) on "German Higher Criticism," as we discuss who wrote the early parts of Genesis. Many Christians through the centuries have simply accepted that Moses wrote the book of Genesis, especially before late 19th century "German higher criticism," and many still do today. But what about those powerful statements in the early part of the book, like "And God said," and "And God saw the light," and "And God called" or "And God created"? I mean, how do we know? Who was there?

Moses *wasn't there* at the beginning— *he came along over 2,000 years later*—but he was unquestionably the key protagonist and likely author of the other four books of the Pentateuch. But the repeated lead-in, *"These are the generations* (toledoth) *of . . . "* is the interesting recurring phrase in Genesis, seemingly dividing the book into sections, and pointing to actual authorship. It seems logical that Moses was not the true *original* author of Genesis, but the secretary, overseer, and/or editor of a series of earlier historical documents (or tablets), actually written by Adam and the other succeeding patriarchs. Adam and Eve would have been created fully formed by God, with a perfect untainted-by-mutation (not yet fallen) brain and body, with reasoning, speech and language skills and the ability to walk and converse with God. Consider this:

> It is reasonable that Adam and his descendants knew how to write, and therefore kept records of their own times (note the [*signature*] mention of "the book of the generations of Adam" in Genesis 5:1). These records . . . were possibly handed down from father to son in the line of the God-fearing

patriarchs until they were finally acquired by Moses when he led the children of Israel out of Egypt. During the wilderness wanderings, Moses [apparently] compiled them into the book of Genesis, adding his own explanatory editorial comments where needed. Genesis is still properly considered as one of the books of Moses, since its present form is due to him, but it really records the eye-witness records of these primeval histories, as written originally by Adam, Noah, Shem, Isaac, Jacob and other ancient patriarchs.

It is interesting to note, as an indirect confirmation of this concept of Genesis authorship, that while Genesis is cited at least 200 times in the New Testament, Moses himself is **never noted** as the author of any of these citations. On the other hand, he is listed at least 40 times in reference to citations from the other four books of the Pentateuch. There are also frequent references to Moses in the later books of the Old Testament, but **never** in relation to the book of Genesis.[198]

At this point a reminder is in order of *theopneustos*, which means *God-breathed*. II Timothy 3:16-17 states that "All Scripture is given by inspiration of God (one word in the Greek: theopneustos), and is profitable for doctrine, for reproof, for correction, for instruction in righteousness: That the man of God may be perfect, thoroughly furnished unto all good works." That is, *all* of the words (not just the thoughts or ideas) are inspired of God. Dr. Morris continues on to logically finish with a powerful, thought provoking conclusion:

In sum, we can be absolutely confident that the events described in Genesis are not merely ancient legends or religious allegories, but the actual eyewitness accounts of the places, events and people of those early days of earth history, written by men who were there, then transmitted

---

[198] Henry M. Morris, Ph.D., LL.D., Litt.D., annotations and appendices, *The Defender's Study Bible*. World Bible Publishers, Inc., 1995, p. 2

down to Moses, who finally compiled and edited them into a permanent record of those ancient times.[199]

In closing this section, I'll admit to many compromising seminary professors, pastors, and theologians who not only have been greatly influenced by ungrounded evolutionary/long-ages dogma, but whom likely would disagree with this view of Genesis authorship. Along with white lab coat intimidation of science teachers and academics, that of *pastors and seminary professors* is unquestionably one of the main reasons we're in the cultural mess we're in today. Even many of the modern "versions" of the Bible came about for related reasons, or had at a minimum certain words or terms that were eliminated or subtly changed. For example, the relatively recent but widely-used New International Version (NIV) has some 64,000 words removed from the Authorized (or King James) Version; that's one out of every eleven words! But, alas, that topic awaits a different book.

---

[199] Ibid, Morris, *Defender's* . . . p. 2

42

WHERE ARE THE GRAVES?

THE WONDERFUL, EASY-TO-READ book *Inspired Evidence* discusses many examples from science and history that support the Bible and the historical narrative of Genesis as written. One important but infrequently discussed subject the authors touch on, relative to the real history of mankind, is the nature and number of human remains found in graves and burial sites. Evidence and history show that humans have always buried their dead. This includes even so-called Neanderthal and Cro-Magnon man, people alive shortly after the Flood of Noah who found shelter in caves, who buried their dead along with valuable tools and artifacts. Besides pointing to skill and intelligence, this **indicates two important things** about human history. Here is the authors' logical and compelling argument:

> First, people have always had knowledge that this life is not all there is. From the ancient Egyptians and Chinese to pre-Columbian cultures right up to recent so-called *Stone Age*-like tribes in New Guinea, humans have prepared their loved ones for life after death by burying them along with treasured possessions. Secondly, humans could not possibly have been around for the often touted 50,000 or more years. Suppose: 1) For most of those purported 50,000 years the entire world population averaged only half a million humans, and 2) These "primitive" humans had an average life span of 50 years. Even using these *extremely conservative* numbers there should be 500 MILLION Stone Age graves out there someplace! Yet only a few thousand are known to exist. This is simply because mankind has been on this planet for only several thousand years. Stone Age people were simply those few living in caves during the ice age, which followed the flood of Noah. We can

understand why so few graves are found when we look to the
Bible as the true history of the world.[200]

The idea of primitive ape-like humans, intermediate between apes and
modern man, is ridiculous, unsupported by archaeological, paleontolog-
ical, and genetic findings, and the desperate race to find human "missing
links" is filled with outright fraud and deception. Notwithstanding academic evolution-based indoctrination, mankind's history—including origin
of the stunningly advanced most ancient civilizations and the various languages—go back only 4,300 years or so (to right after the Flood of Noah,
and the Confusion of Tongues at Babel). The paucity of human graves is
easy to understand; just put your biblical glasses on.

---

[200] Ibid, Von Vett and Malone, *Inspired* . . . April 7th history reading

43

PSYCH DOCS, THE POOR LIMEYS, AND NO-HOPE KIDS

PSYCH DOCTORS—PSYCHOLOGISTS AND psychiatrists— who are the experts that offer a veneer of professional or pseudo-scientific research or validation for the slam on historic American manhood (toxic masculinity), but obvious support for transgenderism, gender fluidity and interchangeability, and the homosexual lobby, have a big Convention coming up (Aug., 2019). One of the highlighted continuing education workshops to be offered: *Leaning Into Masculinity: Applying the New* APA (American Psychological Association) *Guidelines for Psychological Practice With Boys and Men.* Under course description, it says this:

> Psychologists are now re-thinking masculinity through the same multicultural lens that feminist researchers have utilized in the study of women.

Uh-oh, we're in big trouble, guys. This is the group that reclassified homosexuality from a *mental disorder* to merely an alternative lifestyle in 1973, and now they're going to adopt the multicultural perspective of leftist feminist researchers?! Be forewarned of a high gobbledygook quotient. This is basically a group of psych *experts* who after the last four years or so of *research* and chit-chat came up with the conclusion that traditional masculinity is harmful. This is totally ungrounded and agenda-driven, but nevertheless will be embraced by the feminists and homosexual lobby to push their public school policy proposals and preferences, and exert more influence on everything from administrators to library book selections to guidance counselor philosophy, in their attack on traditional masculinity and biblical roles for men.

Another upcoming continuing ed course at the August Convention: 123: *TRANSlating Research into Practice*: *Evidence-based care for the Transgender Community*, and mentions as part of the program discussion cutting-edge research regarding the use of psychological assistance with gender diverse individuals. In my experience and in this kind of context, *cutting edge research* usually means secular or far-left agenda, lifestyle, and/or worldview promotion. One of the objectives in this course: 'Evaluate current federal policies and legal actions that will enable you to serve as knowledgeable and effective *advocates for transgender persons*.' It is just mind-boggling where we are in this culture today. Recently a professor in Madison, Wisconsin urged schools to implement toxic *masculinity* training in kindergarten. Don't be intimidated by the appearance of professional endorsement, bloviation by academics with lots of letters after their name, or promotion by the mainstream media. Do your homework, seek truth, stand your ground for right and beware of the terminology, talking points, and deceptive words—like *tolerance*—one of the left's favorites. Do not be deceived.

As mentioned before, not only do higher academia and the fake news media downplay the uniqueness and exceptionalism of America, they blame our country for many of the world's problems. From this point on, I'll no longer use the term "mainstream" media, because they're no longer mainstream, as their bias, dishonesty, and left-leaning worldview is so extreme and obvious. The New York Times recently termed this American so-called 'greatness' simply a *myth,* right before the July 4th celebrations; we're just . . . "okay," at best (and beware of the Times' pitiful, revisionist history *1619 Project*). It's no wonder our young people are so confused, with all of the rewritten history, lies, indoctrination, climate change propaganda, and censorship. The now obvious goal of *invalidation of America's Founding*, and the radical transformation of the United States from its Judeo-Christian ethic—some would say the destruction of the country as we know it—is being done primarily through the souls of our youth.

Consider our kids and young adults, and think about what they regularly hear, and what they're taught: They have evolved from the primeval soup; there is no God; America is just an imperialistic slime-ball country; and climate change is going to destroy us. We've got glaciers and the poles melting while the poor polar bears and musk oxen struggle, and our coastal cities are going to be underwater soon (how did Al Gore's prediction on that pan out?!). Our young people are taught not only the acceptance and promotion of the transgender, lesbian, and homosexual lifestyle and gender fluidity (even in the early grades), but in some quarters its *preference*. The *age of innocence* is virtually a thing of the past.

They are also alarmed that coal, gasoline, oil and nuclear fuel use is going to decimate us and the environment; that *hate* crimes are supposedly everywhere- especially by whites (blatantly false); divisive *identity politics* is the way to go; and our American heroes' monuments and statues need to come down; and on and on . . . so they essentially have **nothing to live**

for. This goes hand-in-hand with an all-encompassing, materialistic, *unscientific* billions-of-years evolutionary worldview, where they are taught that they're just another animal, merely a biological accident, and thus there are no life-enriching unchanging moral laws. The final no-hope, life-sapping crescendo, to cap it all off: when you die "you simply rot away in the ground"—as famous English atheist and evolutionist Richard Dawkins likes to say— and "that is the insignificant end of your meaningless life."

No wonder the English now have so many problems. Speaking of them, and along this same thread, British Prime Minister Teresa May recently said that "loneliness is the sad reality of modern life," while announcing the new position, *Minister of Loneliness*. I kid you not! This reminded me of another bureaucratic entity the Brits created some years ago called the *Office of Redundant Churches*, because almost nobody was going to church anymore over there in the land of Darwin and Dawkins, and they had to decide what to do with all of the ageing and empty church buildings. Many became Islamic mosques. The current mayor of London is even a Muslim—pretty amazing. My perception is that he's a decent guy, but there are now Muslim neighborhoods in some of England's larger cities that whites and the police do not even venture into. Remember, we're talking about what was probably the most powerful country in the world 150 years ago. England is just a shadow of its former self, as it has cast aside God in the last century.

Back here across the pond, it's as if the very ethos of our American culture has disturbingly changed, as biblical authority has unjustifiably waned and we've turned down the same road as Great Britain and other European countries. Is it any wonder that suicide is on the big increase, our young adults are called the loneliest generation *ever*, we have a huge drug problem, we've cast aside our social mores, mass shootings abound, and all with declining American manhood and confused, unconfident men everywhere? Everyone who loves this imperfect but good country and righteous liberty needs to work to get America back on the right road, and beware of psych doc endorsements. Perhaps the most frustrating, perplexing observation related to the entire problem: the United States of America is the only country in the world where the vitriolic secular critics and leftist social justice warriors, who claim to absolutely *hate* it, absolutely *refuse to leave it*!

44

WANT YOUR FACE TO *SHINE*? HOW ABOUT SOME "FEAR" OF THE LORD?

IN THIS CONFUSED self/sexuality-focused culture guided by tolerance and political correctness where everything is relative, have you noticed that the strident social justice warriors, radical feminists, and left-leaning media pundits are never happy and *never smiling*? It is so undeniably obvious. Could that possibly be related to knowledge and wisdom, or *the lack* thereof? In Proverbs 1:7 it is stated that "The *fear* of the Lord is the beginning of knowledge: bur fools despise wisdom and instruction." Socrates probably would have disagreed; he said "to know thyself is the basis of all knowledge" (he likely snagged that one from the inscription on the Temple of Delphi).

Well, I love learning and desire knowledge, and I would think every serious truth-seeker feels likewise. I would say knowing yourself, and being honest about it—especially your strengths and weaknesses—is quite important, but I'm siding with Solomon instead of Socrates on this one and putting belief in the Lord, with awe and reverential respect (*fear*), at the top of the list. In Psalms 111:10 and Proverbs 9:10 it proclaims "The *fear* of the Lord is the beginning of wisdom." The main theme of the book of Proverbs is to know true wisdom, and its foundational "seven pillars" of Christian character are also found in the New Testament in James 3:17. In Job 28:28, it says "Behold, the *fear* of the Lord, that is wisdom; and to depart from evil is understanding," and then in Job 37:23-24, it says the Almighty "is excellent in power, and in judgement, and in plenty of justice: he will not afflict. Men do therefore *fear* him." Also in Solomon's advice-for-living Book of Proverbs, in verse 10:27 it states "The *fear* of the Lord prolongeth days." Following that in 16:6 is found another example that says "and by the *fear* of the Lord men depart from evil"; further on it says "the *fear* of the Lord is clean." In Psalm 103:11, reassuringly to believers, it states "For as the heaven is high above the earth, so great is his mercy toward them that *fear* him." A few examples from the New Testament include Hebrews 11: 7, where "Noah moved with *fear*," and Hebrews 12:28, where it says "let us have grace, whereby we may serve God acceptably with reverence and godly *fear.*"

Listening to an impressive nationally known speaker decades ago—who was also a very successful businessman—he touched on this subject while discussing some of the deleterious cultural changes in America. His bold claim, on what America really needed: *more people to fear God.* I was puzzled and had never heard anybody talk like that, especially outside of church. I wholeheartedly agree with him now. This idea, "the *fear* of the Lord," is used elsewhere in Scripture as well; what does that actually mean?

Having "the *fear* of the Lord" means belief or faith in God with reverential respect, usually followed by obedience, and hopefully an increase in wisdom and some or all of its six permutations: knowledge, understanding, instruction, learning, discretion and wise councils (see Proverbs). Sincere belief in Jesus and the Christian Gospel (with a proper *fear* of God) importantly includes belief in the reality of eventual judgement, often referred to as the judgement seat of Christ (for believers only), where they will be judged not for salvation, but for rewards, or loss of rewards (see Romans 14:10-12; 1st Corinthians 3:13-15; 2nd Corinthians 5:10; and Revelation 22: 12). According to Scripture, *non*believers (atheists), will be judged also by their works, but at the great white throne judgement (Revelation 20: 11-14).

Scripture says emphatically that *the fool* is someone who says there is no God, and that such a person lacks true wisdom (which is more precious than gold, silver, and rubies; see Proverbs 8). I was a fool for many years, because of white lab coat intimidation and evolutionary indoctrination. The Bible also says such nonbelievers "hold (or suppress) the truth in unrighteousness," "willingly are ignorant," and have no excuse for their unbelief, because the evidence for the Creator-God is all around them (Romans 1:18-20). Just a small portion of that evidence is touched on in this book.

On the other side, a sharp young person (or even an oldster) with that real-deal, heart-changing faith (belief) in God—with that reverential awe and respect, or fear, of the Lord—can gain some **serious understanding and wisdom**. In fact, sometimes such a grounded *student* even gets **more** of it than "all of [their] teachers" or professors (Psalms 119:99). Let's go back to so many of the of the humanistic, seemingly always unhappy, *un*smiling social justice warriors. The sage Solomon said "a man's wisdom maketh his face *to shine*, and the *boldness* of his face shall be changed" (Ecclesiastes 8:1). Mark Twain once described a particular self-assured fellow he was dealing with as "having the calm confidence of a Christian with four aces" (that dude probably had a tough time keeping his poker face, suppressing the shine). How about you? Are you confident in your long-term future? Do other people like being around you? Do you want your face to shine, maybe with even a little hint of *boldness*?

The key: first get your heart right, then work on that faith—or belief (same word in the Greek)—and gain some bona fide, Godly, reverential awe and respect (or *fear*). Then eventually utilize that valuable, precious wisdom . . . which is "better than rubies, and all other desired things" (Proverbs 8:11).

45

THE IMAGE-BEARERS

THE LATIN *IMAGO Dei* is a theological term meaning "the image of God," applied specifically to human beings, a Christian concept taken from Genesis 1:27, wherein "God created man *in his own image*." The Christian belief as humans being the only organisms created by God "in his own image," implies not only uniqueness but extreme *value—*or *worth—* of *every* human being. This novel, culture-changing Christian concept transformed the Roman Empire and led to the eventual establishment of entities like hospitals, and early versions of old-folks homes and orphanages. Contrasting with other created organisms, trees and plants have a body, or structure, and the animals also have a body, or physical form—but also a *consciousness*. Mankind goes beyond that, more like the triune Godhead. Human beings, in addition to the body and consciousness, have the image of God, the most important part of which is *an eternal spirit*, which has the potential to worship and fellowship with the Creator God.

Another interesting consideration are the unique *muscles of facial expression* in human beings, which help convey a myriad of different emotional feelings, messages, and expressions, along with a stunningly complex brain and muscular tongue obviously designed for high level speech and language (unique to human beings). Also anatomically unique and important for this is a large throat allowing a variety of sounds (at least fifty), in contrast to that of an ape's, which is shallow. Humans have nearly a hundred muscles for precise control of the vocal cords and speech, while apes have far less, and apes do not have fine finesse control of their tongue and lips, thus greatly limiting the sounds they can make. Apes and other animals are not only incapable of written language and articulate symbolic speech, but they cannot create and appreciate music, admire beauty, or manifest a moral conscience. The Creator of the universe who designed ants, apple trees, and apes designed human beings—*in his own image—*to verbally communicate through language to each other and to Him.

Humans also have imagination, which is part of the image of God. Study of God's specific revealed written revelation (the Bible) clearly shows that God has strong emotions, and he *imagines*. This imagination allows man to be a little creative (a bit like God, but not with *ex nihilo* ability), and to grow, make, and build things from the resources and materials of the earth (under the dominion mandate, Gen. 1:28). Also, related to *"his image,"* is the opportunity *to choose,* or to *make choices.* Thus, humans are not forced, or coerced to believe, obey, and follow God. We are not robots. If that were the case, there could be no choices, no real image-similarity, and more importantly—*no love*. God's created amazing earth with its stunningly complex life forms, along with the majestic heavens and billions of galaxies are indeed awe-inspiring, but the Bible indicates God's main goal—*his primary focus*—is to create and redeem men and women *in his image*. He cares more about *you* than any of that other stuff . . . Wow!

PART V

CLOSING

TODAY'S CULTURAL CONFUSION and ever-deepening rift in America's social fabric is in large part due to the adoption of the ungrounded foundational assumption of evolutionary materialism, the concomitant elimination of moral absolutes and biblical authority, and its resultant no-hope worldview. An attempt is made in this book to not only inform the reader of the lack of scientific support for the evolutionary idea, but also the often unrecognized, all-pervading influence and reach of the evolutionary tentacles, and the iniquitous and sometimes horrific results thereof. As theistic evolutionist and compromising Jesuit priest Pierre Teilhard de Chardin once stated: "[Evolution] is a general postulate to which all theories, all hypotheses, all systems must henceforward bow and which they must satisfy in order to be thinkable and true. Evolution is a light which illuminates all facts, a trajectory which all lines of thought must follow."[201] Influential evolutionist and former Harvard zoologist and ornithologist Ernst Mayr (who passed away in 2005, at age 100), was an intelligent young man who was powerfully impacted by the dominant secular evolutionary worldview, while growing up in post-WWI and then Hitler-dominated Nazi Germany. Similar to de Chardin, in strongly referring to the universal influence and application of evolution, he reverentially wrote the following:

> Man's world view today is dominated by the knowledge that the universe, the stars, the earth and all living things have evolved through a long history that was *not* foreordained or programmed. I am taking a new look at the Darwinian revolution of 1859, perhaps *the most fundamental of all* revolutions in the history of mankind. It not only eliminated man's anthropocentrism, but affected **every** metaphysical and ethical concept, if consistently applied.[202]

---

[201] Pierre Teilhard de Chardin, as cited by Francisco Ayala in "'Nothing in Biology Makes Sense Except in the Light of Evolution': Theodosius Dobzhansky, 1900-1975," *Journal of Heredity* (1977), Vol. 68, #3, p.3

[202] Ernst Mayr, "The Nature of the Darwinian Revolution," *Science*, (June 2, 1972), #176, p. 981

Reading and responding to scientifically unsupported elitist statements like these can be frustrating, with a lot of head shaking. It's pretty doggone profound what you sometimes hear filtering down from the rarefied air of academia and professional punditry—at least Mayr only lied in his *first* sentence (with use of the word knowledge instead of *story*, or *fairy tale*)—but both sweeping claims are totally unsupported by scientific evidence and in opposition to the foundational laws of science. There is **no** illustration of—or even *realistic conception of*—a genetic and biochemical mechanism for the addition or input of *new* genetic information to the genome, which is irrefutably required for onward and upward evolution. There is **no solid evidence** that things are millions or billions of years old, which is absolutely required for the evolution idea to be even *remotely* feasible. Macroevolution has **never** been observed in the **present**—in nature or the laboratory—**or in the past or fossil record** (with instead *stasis*, and its stark absence of transitional fossil forms). Despite over a century of media, academic, and Hollywood pseudo-scientific indoctrination, it can justifiably be stated that it **is not even possible** because of the aforementioned universal laws of conservation and decay that now govern all natural processes, and seemingly the entire universe (the 1st and 2nd Laws of Thermodynamics). Atheists and evolutionism advocates continue to rail and argue as they frantically search for support for their idea, while as you read these words the translocation of their 'theory' toward its funeral pyre has already been initiated.

This was even indirectly predicted to eventually happen by Thomas Kuhn in his erudite 1962 classic, *The Structure of Scientific Revolutions*. I suppose the shouting and usual ad hominem attacks are okay—expected, for sure—and everyone is entitled to their own opinion, but they *are not entitled anymore to their own set of facts*. Most dating methods point to a very young age for the earth, and the few that sometimes indicate great age have been proven unreliable, and are based on questionable and unprovable assumptions about the past, including the denial of a relatively recent global Flood/tectonic cataclysm. Such an event would **totally invalidate** such long-ages radiometric dating methods. This book also exposes widespread bias and censorship in our schools, textbooks, science journals, media and museums by evolution *experts* and proselytes, and false claims and frauds. Exposed also is the enthusiastic and quick adoption of the evolution paradigm by academic leaders, wealthy influential industrialists and robber barons, and societal elites (in addition to the two just quoted) a century or more ago, and the resultant horrific, worldwide cultural and societal ramifications (bad fruit) from applied Darwinism. I prefer the word *story* when discussing the evolution idea, because it does not come close to qualifying as a real scientific theory, nor, some would say, even as a hypothesis. It is truly and simply *just a story* to explain things without God. But do not underestimate its influence:

> As far as Christianity was concerned, the advent of the theory of evolution and the elimination of traditional theological thinking was *catastrophic*. The suggestion that life and man are the result of chance is incompatible with the biblical assertion of their being the direct result of intelligent creative activity. Despite the attempt by liberal theology to disguise the point, the fact is that no biologically derived religion can really be compromised with the fundamental assertion of Darwinian theory. Chance and design are antithetical concepts, and the *decline in religious belief* can probably be attributed more to the propagation and advocacy by the intellectual and scientific community of the Darwinian version of evolution *than to any other single factor.*[203]

Well over a half century ago American historian Will Durant, who along with his wife (and historian partner) Ariel was strongly influenced by Darwin's godless evolutionary idea in the early 20th century, offered a similar perspective from a different angle: "The greatest question of our time is not communism vs individualism, nor Europe vs America, not even the East vs the West; it is whether men can bear to live without God."

Hard-working but frustrated citizen-patriots, who respect the Declaration of Independence and Constitution and cherish family, fairness, American values and belief in a Creator-God (that the left now mocks and denigrates), must get a grasp of this. It is absolutely vital if we want to begin to comprehend *how we got in the mess we're in*, and collectively try to make effective, grass roots, positive differences in our individual spheres of influence. This may be the only way to save the America we respect and love (warts and all). America's downward cultural spiral, palpable national angst, destroyed young lives, and unjustified disrespect of biblical authority need to be recognized and rectified. This book is partially a wake-up call, by getting some heretofore censored information and truth in science and history out there, to help save America from secular

---

[203] Michael Denton, *Evolution-A Theory in Crisis*, Burnett Books, London, 1985, p. 66. Dr. Denton was an intellectually honest biochemist/genetic researcher and evolutionist when penning this book 35 years ago. Always insightful and fortunately still intellectually honest, he presently is a proponent of intelligent design, and a Senior Fellow at the Discovery Institute's Center Science and Culture.

academic and entertainment elites, activist judges, a corrupt mainstream media, and dysfunctional self-aggrandizing politicians. It is also a needed exposure of the unjustified attack on men and manhood.

Just complaining, calling, emailing, or writing your Representative or Senator inside the Beltway, and voting for the best candidate is now obviously not enough. It's a challenge for sure, particularly when the secular mainstream purveyors of information are now indisputably dishonest and agenda-driven. Real, objective journalism is now almost completely out the window.

Hopefully this book will serve not only as a culture/science/history truth handbook, but as a stimulus or call to action. Contrary to claims of the left, there *is* real truth. The great Winston Churchill stated boldly: "Truth is incontrovertible, ignorance can deride it, panic may resent it, malice may destroy it, but there it is." Born in the same year as Churchill (1874), compatriot G. K. Chesterton put it this way: "You have the right to believe anything you want, but the truth is still the truth." A nation-state without solid truth standards and the rule of law is in trouble. A country can lose not only the "ancient landmarks" but its *freedom* in just one or two generations. Truth principles need to be embraced, taught, and repeated over and over. This is absent from public education today, even in Civics classes (where they might still exist). It takes action, commitment, and guts—*an investment*— **freedom is not free**. Concerned parents, all across the fruited plain, need to retake the high ground in directing and overseeing their children's education.

Perhaps the words of President Ronald Reagan should be heeded: "I do not believe in a fate that will fall on us no matter what we do. I do believe in a fate that will fall on us if we *do nothing*." How encouraging and effective it would be if Americans everywhere increasingly step up, get informed, get involved, and stand for what's *right*. I sense more such activity recently (late 2019) as a result of the unbelievable, embarrassing, criminal, made-up, no evidence, millions of taxpayer dollars wasted, bogus and corrupt mainstream media/FBI/DOJ/Democrat Party scheme to oust President Donald Trump, including the sham impeachment in the Democrat-controlled House of Representatives. What happens in these situations when those in power disregard the law—or illegally weaponize their position—and any residual hint of a moral compass disappears? The answer: they simply operate as Mr. Lenin, Mr. Stalin, Chairman Mao, and Mr. Hitler did; *the end justifies the means*. Consider what eventually happens to a nation whenever the divine lawgiver—the moral compass—is eliminated. Who or what actually determines then what's right or wrong? Who sets the rules? The answer: simply look at history's examples, like the four just-mentioned despots: *Whoever has the power, whoever has the guns.*

In opposition to unlimited *tolerance*, identity politics, relative morality, revisionist history and the attack on and removal of our founding documents, monuments, and ancient landmarks, a solid return to our Judeo-Christian foundation is in order. It is a monumental challenge but an absolute necessity for maintaining true freedom and liberty (not licentiousness-which destroys liberty) and keeping our country. In 1831, the French diplomat, sociologist, and political theorist Alexis de Tocqueville traveled to the young United States with his friend and fellow magistrate Gustave de Beaumont, partially to study its prisons and sociology, but more specifically to potentially collaborate on a larger endeavor to observe "what a great republic is."[204] The following well-known quote is often attributed to de Tocqueville, but I personally have never been able to actually find it, even in his famous (and lengthy) political and social commentary, *Democracy in America*, which sprang forth from his 1831 trip to the still fledgling but energized and growing United States. Nevertheless, the spirit of the quote seems to match the essence of de Tocqueville's feelings about the young, and what he called the "great" American republic:

> I sought for the greatness and genius of America in her commodious harbors and her ample rivers-and it was not there . . . in her fertile fields and boundless forests-and it was not there . . . in her rich mines and her vast world commerce-and it was not there . . . in her democratic Congress and her matchless Constitution-and it was not there. Not until I went into the churches of America and heard her pulpits flame with righteousness did I understand the secret of her genius and power. America is great because she is good, and if America ever ceases to be good, she will cease to be great.

Regardless if that famous quote was the scholarly Frenchman's exact words, he also logically discusses the challenge of *a republic* and especially as it becomes large, diverse, and powerful (like today's USA):

> What one can say with certitude is that the existence of a great republic will always be infinitely more exposed [to peril] than

---

[204] Andre Jardin, *Tocqueville: A Biography*, L. Davis trans., New York: Farrar Straus Giroux, 1988, p. 90

that of a small one[205]. . . . All the passions fatal to republics grow with the extent of the territory, whereas the *virtues* that serve as their support *do not increase* in the same measure . . . Great wealth and profound miseries, metropolises, depravity of mores, individual selfishness, [and]complication of interests are so many perils that almost always arise from the greatness of the state.[206]

My goodness, was he somehow momentarily translated into the future to early 21st century America?! Mr. de Tocqueville also sagely describes how tyranny can spread, often initially to the political world but then "penetrates private life . . . [and then] dictates taste; after the state [infiltration and control], it wants to govern families."[207] I hope you readers understand it does not have to be just a military coup-type tyranny. When virtue is lost—as Washington, Adams, and Madison worried about—it can sometimes take on a different, more modern, and even *more insidious* form. We're witnessing it all around us now with political correctness, media and academic censorship and bias, corruption by the now undeniable *deep state*, the suppression of free speech (especially on college campuses), strident pushers of identity politics, and the politicization of children along with the aggressive political agenda of the radical feminists and homosexualists. Many citizens still do not realize how the demand for sexual freedom is being politicized by these radical ideologies (their comfy "safe places" and headquarters being found in higher academia), and exploited by governments (particularly that of the United States), whose operatives and minions are able to greatly increase their power as a result. Author/political scientist Stephen Baskerville discusses this huge, but poorly perceived problem on the opening page and throughout his powerful and highly recommended book, *The New Politics of Sex*:

One can welcome this development or deplore it, but there can be no doubt that it is taking place, even though scholars and journalists across the spectrum have determined to avert their eyes. . . .A major reason is *fear*. One feature of sexual politics is the almost **complete absence of critical**

[205] Alexis de Tocqueville, *Democracy in America*, Harvey C. Mansfield and Delba Winthrop tran's & ed's, University of Chicago Press, 2000, p. 150-151

[206] Ibid, p. 151

[207] Ibid, p. 150

scholarship that approaches it from any viewpoint other than enthusiastic advocacy. Ostensibly objective scholars are often active participants and promoters of the phenomenon they should be studying and understanding critically. Scholars who refrain from endorsing sexual liberation and insist on analyzing these subjects from a detached perspective find it almost impossible to publish their work and **are quickly driven from the universities**. (*Note: biology, geology, and other science professors and teachers who challenge the evolution/millions of years paradigm—even backed up with sound evidence and research studies—face the same thing. These two issues are *not* unrelated-*RJ*). . . .The fact is that **the western academic world today is *not* an "open society" of inquiry and critical thinking. It is largely closed, inbred, and controlled by heresy-hunters who vet scholarship according to a litmus test of political doctrine and punish heterodoxy with ostracism.**[208]

I myself have spent much time in and around universities, fortunately less so in the last few years. Our freedom to think for one's-self is now under assault. We've got a problem there of a magnitude that would stun many Americans, and one that has contributed mightily to our cultural discombobulation and national angst. Is the promotion of this atheistic, radical sexual agenda and worldview *really* a serious concern? Just simply look around, watch the news, and tolerate repetition of the late judge, legal scholar, and former Solicitor General Robert Bork, and this from a 25 years ago (it's even more true today): "Of all the radical ideologies that seek to control the institutions of learning today, "feminism" is by far the strongest and most imperialistic, its influence suffusing the most traditional academic departments and university administrations."[209] Have you wondered in recent decades about some of these new disciplines, departments, and majors at universities? Again, the actual reality is described well in *The New Politics of Sex*:

Like all ideologies, sexual radicalism is spread by cadres of academic ideologues. Novel disciplines like "women's studies," "gender studies," and "queer studies" recast all knowledge as

---

[208] Ibid, Bask, pp. 1-3

[209] Ibid, Bask, p. 3, taken from Robert H. Bork, *Slouching Towards Gomorrah*, chapter 11, pp.193-225, excerpted at http://fathersforlife.org/feminism/borkch11.htm

sexual-political grievances, and sexual activists have colonized other disciplines, where they exert a veto power over what others may write and say. This is true foremost in the social sciences and humanities but even extends to the natural sciences.[210]

Readers and parents are encouraged to find out what is really going on at colleges and universities of interest, not to mention kindergartens, elementary schools, middle schools, high schools and even private schools where you may be investing some money. In many cases critical thinking, civil debate, honest evidence-based history and science teaching has been thrown in the dumpster.

As mentioned earlier, it has been said by more than one sage that for a good, productive life with contentment and inner joy one needs something to do, someone to love, and something to hope for. Is not *hope* the best (sometimes the last), and most important? Without it, there is only . . . what? Just . . . time . . . or, simply nothingness? Or, perhaps what journalist-poet Ambrose Bierce once referred to as *oblivion*? In his century-old *Devil's Dictionary* he defined it as "the state or condition in which the wicked cease from struggling and the dreary are at rest," and "cold storage for high hopes." Real-deal Christ-followers believe that the physical death sentence we all experience is an *intruder*— the "last enemy" (I Cor. 15:25). It is something we brought on ourselves—"by one man"— or the "first man, Adam" (Genesis 2; I Corinthians 15:45). This is stated repeatedly and emphatically in the Bible (see the all-important passage from Romans 5:12-21), and is actually backed up not only by real history but now also indirectly by scientific evidence, including genetic entropy and mitochondrial DNA studies.

Christians believe physical death is just a step to the glorious, eternal life with our Creator and Redeemer, who has a holy hatred of death. Gracefully and gloriously through the *"second man"*—or "last Adam" (I Cor. 15:45- 47), Jesus Christ—*hope abounds* and things can be made right. These scriptures affirm not only that Adam was, in fact, the first actual man, refuting any long ages evolutionary ideas with imaginary pre-Adamite men, but show Adam to be a contrasting type of Christ. Dr. Henry Morris sums it up well:

---

[210] Ibid, Bask, p. 3, and natural science allusion refers to "Feds Paid $709,000 to Academic Who Sudies How Glaciers Are Sexist," Daily Caller, 7 Mar., 2016,http://dailycaller.com/2016/03/07/feds-paid—709000-to-academic-who-studies-how-glaciers—are-sexist/#ixzz.42ddMacvj

Both were true men, yet their bodies were formed directly by God without genetic inheritance from human parents. Adam was the first man made a living soul, the federal head of the human race; the Lord Jesus was the first begotten from the dead, the captain of our salvation, the first man made a life-giving spirit. Adam brought sin and death into the world; Christ brought everlasting righteousness and eternal life.[211]

Conversely, evolutionists believe death is a *good* thing, and through millions of years of supposed random mutational changes—with struggle, starvation, disease, predation, bloodshed, and death in nature where "all is red in tooth and claw"[212]—man somehow arose and appeared . . . by random chance. It is still stunning **how different and opposed** these two worldviews are, especially considering the fact that *we all have the same evidence.* Let's look at it in simple chart form, to illustrate this regrettable history, science, and worldview clash:

EVOLUTION →→	DEATH →→	→→ MAN
CREATION →→	MAN →→	→→ DEATH

The *chance*, or randomness factor has always bugged me. It is so preposterous and unrealistic, as information *never* comes from random-ness. Even Albert Einstein, as he dove deeper into studies of the atom and quantum mechanics, and the unseen but undeniable nebulous forces at play, was bothered by the uncertainty of it all, and said "God does not play dice with the universe." In contemplating the order, balance, invisible maintenance forces and majesty therein he also added: "the mathemat-ical precision of the universe reveals the mathematical mind of God." That statement, of course, would disqualify Mr. Einstein from teaching in public schools and most universities today.

Atheists and many evolutionists believe mankind does indeed face such dark oblivion, or nothingness . . . other than probably becoming eventual worm food . . . with *no hope.* British-born atheist turned agnostic philosopher,

[211] Ibid, Morris, ed., *Defenders*, p. 1275

[212] A phrase popularized by English poet Alfred Lord Tennyson in the nineteenth century. Tennyson was strongly influenced by lawyer/amateur geologist Charles Lyell (who was a mentor to Charles Darwin), and his *Principles of Geology*, and later by Charles Darwin's famous 1859 *On the Origin of the Species.*

Florida State University professor and evolutionist Michael Ruse (already referred to in Part I), struggles in his new book, *A Meaning to Life* (2019), to come up with a Darwinian meaning for life, as he claims "We come from an eternity of oblivion. We return to an eternity of oblivion." Wow, I wonder if you FSU students and alums were uplifted and energized in *his* classes?! Sometimes a lot of graduate degree letters after one's name can be a stumbling block if on a truth search. First Corinthians 2:14 says this: "For the natural man receiveth *not* the things of the Spirit of God: for they are foolishness unto him: neither can he know them, because they *are spiritually discerned.*" Especially if kids are present, have you ever wondered about the family dynamic in an atheist's household, particularly around Christmas or Easter?

One of evolutionist Richard Dawkins' well-known statements: 'There is no purpose, no evil, and no good. We are merely dancing to our own DNA.' One wonders—with some head shaking— what world is Dawkins living in?! I'll give him credit, though; it is an accurate, honest summary of the evolutionary worldview, and you can't have *evil* without *good*, and you can't have *good* or even substantive *purpose*, without *God*. I remember years ago hearing a Charlotte pastor talking about this challenging life in a fallen world, and he said "Without Jesus Christ, life is just a prison sentence to be endured." That one resonated with me, as I saw many people seemingly living that kind of life. Even the great Leonardo da Vinci felt that a painter should probably start every canvas with a layer of black, because everything in the natural world is dark, except when exposed by light.

Memorable medical maverick Dr. William Campbell Douglass offered a different perspective: "when you get down to the real nitty gritty of reality— life and the Creator—you're talking about *light*, which is believed to be the smallest *particle*. But this *particle* has no mass, and therefore exists only in the mind through the perception of the brain." He goes on to comment:

> If you combine two of these massless *non-existent* light particles, then you have an electron which has mass and weight; you can "pick it up," so to speak. You have passed from the unreal world of quantum mechanics to the material world as we view it. How does light get transformed into physical substance? How do two things become a *some*thing? That's God's little secret, and I doubt he will ever reveal it. There are limits to what man can understand—at least I think so.[213]

---

[213] William Campbell Douglass II, M.D., *Into the Light*, Rhino Publishing, S.A., Panama City, Panama, 2003 edition, p. 1

So here he alludes to one of those most amazing "invisible things" (Romans 1:20), and I'm reminded of Isaiah 55:8-9, and Romans 11:33 where it says God's thoughts and ways are not only much higher than ours, but sometimes "past finding out."

The Bible and history reveal Jesus Christ is not only the Truth and *the light*, but that energizing *hope* . . . and Jesus is *real*. Jesus was born . . . the 1st Advent . . . **to save us**. He came 2,000 years ago *not* to bring the hammer down and judge, but **to *save* you and me**. After the eventual 2nd coming (or 2nd Advent) He *will* judge, but at the first coming, and through His ministry, Crucifixion, Resurrection, Ascension, Pentecost and on through the centuries to the present day with the work of the Holy Spirit, He wants to *save* us. Because God is a *holy* God of love, his standards of justice are high, and our penalty could only be paid by one *sinless and blameless* (similar to the innocent animal God slew to make clothing, or a "covering," for Adam and Eve's sin; Genesis 3:21). What love . . . what a free gift! What a *Gospel*—or "good news"—which makes no sense without the *bad news*. One just needs to *receive* it.

There is something undeniably wrong with those of us "in Adam" (mankind); we are all tainted. We want to reject biblical statutes—or rebel, and do our own thing—particularly apparent in this *post-truth* age of the 21st century. This tendency even included the Lord's *all-in* loving lieutenant— the Apostle Paul. He wrestled with this (what he called the law of sin): "For the good that I would I do not: but the *evil* which I would not, *that I do*" (Romans 7:19). It is obvious that you don't even need to teach a human toddler how to sin; they're *born with* that sin nature. You can put two cute, cuddly little human 13 month-olds in the nicest playpen available, with eight or nine cool toys or playthings, and in no time they'll be fighting over the same one. With blinders on, atheists and humanists and some evolutionists deny this inherent *sin* nature and most still believe—despite mountains of historical evidence to the contrary—that man can fix his own problems, with no need of a divine lawgiver or moral compass.

God sent his Son, our Creator (see John 1:1-3, Ephesians 3:9, Colossians 1:16-17, and Hebrews 1:2-3) who came willingly—the *only* sinless and blameless one available— to be judged (propitiation) *in our place*. That is why it had to be a miraculous conception, and virgin birth, with Mary of *"that holy thing"* (Luke 1:35). Any contributing genetic material from either Joseph *or* Mary (with their *fallen* nature, like ours), would have tainted the embryo, and thus such a child would not have lived a sinless and blameless life. Jesus was still fully human, though, while still fully God.

If God *just wanted to judge* the world, He surely would **not** have sent his Son to go through all of that horrific abuse and travail, temporary separation from the Father, and tortuous crucifixion. He also *wouldn't have given* *us the Book.* Jesus, who could have called in through the Father "more than twelve legions of angels" (Matthew 26:53) at any time (and I'd bet big money those thousands of powerful angelic warriors were just chompin' at the bit to come!), came to offer us the greatest second chance, the most needed mulligan, the most important reprieve shot (for you basketball *horse* players) of all time. And how about the voice of that Jesus, whose "word was *with power*" (Luke 4:32; Luke 4:22). Even his enemies admitted it. The plotting Pharisees were extremely irritated at their henchmen ("officers," John 7:45) who failed to drag Jesus in before them. Their response in the next verse, when angrily questioned why: "The officers answered, 'Never man spake like this man.'" Wow. This is real history, folks. He, who didn't need to, came to save those, who didn't want Him. Most kings and lords get their people to die *for them*; King Jesus did the opposite—He died **for his people**.

Why do Christians want other people to believe? Why am I sharing suppressed truth and expressing that same desire through this book? Why should I care what your personal beliefs are? Respect for biblical authority and the foundational biblical family unit—an institution ordained by God—is under massive attack today. God works through families. Society functions better with a plethora of God-fearing families. I just want to get some truth out there and help right the ship. As already mentioned, I was indoctrinated and misled with ungrounded evolutionary no-hope dogma by teachers and professors—*white lab coat intimidation*—and it negatively impacted my life. There are multiple other reasons—some are discussed in this book— but Christians want people to have hope and a bright *long-term* future, and the peace which passes all understanding. The latter is so lacking in the world today—with conflict, confusion, corruption, strife and snuffed-out dreams seemingly everywhere. As semi-salty North Carolina preacher and evangelist Vance Havner used to say, "This earth is not a playground— it's a battleground!" (Ephesians 6).

We Christians *believe*, and we are *thankful*, that our Creator became our Redeemer. What unfathomable love! Do Christians still make mistakes and have challenges? They absolutely do, sometimes more so. Christianity is *not* filled with perfect or even good people . . . just *forgiven* people. Jesus' closing command was "to go out and teach everything I have commanded you." God has "called us to the ministry . . . and word . . . of reconciliation" (II Corinthians 5:18-19), as "ambassadors" of Christ to share with and beseech people everywhere to *receive* His Son and His amazing work of salvation (often called the *Great Commission*).

The wonderful reality is that God has *already reconciled* sinners to Himself through His unfathomable grace and sacrifice of his Son. God used the worst event in human history—the crucifixion of Jesus Christ, his Son— to gain the salvation of sinners (i.e., the rest of us). The challenge now—and

Christ-followers' loving concern—is that fellow sinners are not yet reconciled to Him. The phenomenal truth is that Jesus, our Creator, **came to *save* you and me**. Years ago I heard a fine pastor say the Holy Spirit *is a gentleman*. He doesn't force Himself on you—that would not be love—it's a *choice*. It doesn't cost you anything; you don't have to pay for it, or earn it. In fact, you can't. You also don't have to make a bunch of changes and spruce up your act first; you can't really do that either, anyway. It's an amazing free gift—come as you are . . . *it's all about grace*. Does *getting a new heart, a new life,* and eternal life with Jesus and the saints appeal to you? You just need to *receive* the free gift.

APPENDICES

APPENDIX A

Here are just a few Great Books and Resources on the *Recent and Rapid* Formation of Fossils, Fossil Fuels, Diamonds, and Petrified Wood. These include Dating Methods and Rates that point to *Young Age* for the Earth. (All are unlikely to be found in Public, Secondary School, or University Libraries because of censorship).

The Fossil Record, by John D. Morris and Frank Sherman, Institute for Creation Research, Dallas, TX, 2010

The Rocks Cry Out, series of 18 DVDs, by Bruce Malone, Search for the Truth, Midland, MI, 2016

Grand Canyon-a different view, Tom Vail, Master Books, Green Forest, AR, 2003

In Six Days, by John F. Ashton, PhD, editor, Master Books, Green Forest, AR, 2000

Science and the Bible, by Henry M. Morris, Moody Publishers, Chicago, IL, 1986

Inspired Evidence, by Julie Von Vett and Bruce Malone, Search for the Truth, Midland, MI, 2017

Rethinking Radiometric Dating, Evidence for a Young Earth from a Nuclear Physicist, by Vernon R. Cupps, Institute for Creation Research, 2019

Evolution Exposed-Earth Science, by Roger Patterson, Answers in Genesis, Hebron, KY, 2008

Censored Science, by Bruce Malone, Search for the Truth, Midland, MI, 2014

Wild, Wild Weather: The Genesis Flood & The Ice Age, DVD, by Dr. Larry Vardiman, Answers in Genesis, Florence, KY, 2003

Recent Rapid Formation of Coal & Oil, DVD, by Dr. Andrew Snelling, Answers in Genesis, Hebron, KY, 2017

A Matter of Time, DVD, by Bruce Malone, Search for the Truth, Midland MI, 2012

Contested Bones, by Christopher Rupe and Dr. John Sanford, FMS Publications, FMS Foundation, 2017

Dinosaurs and the Bible, by Brian Thomas, Institute for Creation Research, Dallas, TX, 2013

Man, Dinosaurs and Mammals Together, (booklet) by John Allen Watson, Mt. Blanco Publishing Co., Crosbyton, TX, 2001

Acts and Facts (outstanding magazine relative to the origins debate, dating methods, and the Culture War), Institute for Creation Research, Dallas, TX; ICR.org/subscriptions, or 800 337 0375

Answers Magazine (outstanding magazine relative to the origins debate, dating methods, and the Culture War), Answers in Genesis, Hebron, KY; see answers-magazine.com

Starlight and Time-*Solving the Puzzle of Distant Starlight in a Young Universe*, by Dr. Russell Humphries, Master Books, Green Forest, AR, 1994 (eleventh printing, 2008)

The Lie: Evolution/Millions of Years, by Ken Ham, Master Books, Green Forest, AR, 1987, 2012

www.answersingenesis.org

www.icr.org

www.searchforthetruth.net

Age of the Earth, Compromises and Crucial Assumptions of "Millions of Years," by Dr. Andrew Snelling, Answers I Genesis, Hebron, KY, 2016

Genetic Entropy, by Dr. John Sanford, FMS Publications, FMS Foundation, 2014

Brilliant-Made in the Image of God, by Bruce Malone, Search for the Truth, Midland, MI, 2014

The Fossil Book, by Gary and Mary Parker, Master Books, Green Forest, AR, 2005

Have You Considered?, by Julie Vonn Vett and Bruce Malone, Search for the Truth, Midland, MI, 2017

Bones of Contention, by Marvin L. Lubenow, Baker Books, Grand Rapids, MI, 1998

The Young Earth, by John D. Morris, Ph.D., Master Books, Green Forest, AR, 1994

A Closer Look at the Evidence, by Richard and Tina Kleiss, Search for the Truth, Midland, MI, 2018

APPENDIX B

A Handful of Important Books on Real History, Real Men and Manhood, Censorship and Free Speech Suppression that may help you to "be ready always," and *Contend and Defend*

The Long War Against God, by Dr. Henry Morris, Master Books, Green Forest, AR, 2002

The Patriot's Handbook, by George Grant, Ph.D., Highland Books, Elkton, MD, 1996

Seven Men Who Rule the World from the Grave, by Dave Breese, Moody Press, Chicago, IL, 1990

R. E. Lee, Volumes I-IV, by Douglas Southall Freeman, Charles Scribner's Sons, New York, NY, 1934-1935

The Great Turning Point, by Dr. Terry Mortenson, Master Books, Green Forest, AR, 2004

The Closing of the American Mind, by Allan Bloom, Simon & Schuster, New York, NY, 1987

In the Hands of Providence, by Alice Rains Trulock, The University of North Carolina Press, Chapel Hill, N.C. and London, 1992 (biography of Joshua Lawrence Chamberlain)

Glass House-Shattering the *Myth* (Theory) of Evolution, by Ken Ham and Bodie Hodge (general editors), Master Books, Green Forest, AR, 2018

Stonewall Jackson: The Man, The Soldier, The Legend, by James I. Robertson, Jr., Cumberland House, Nashville, TN, 1997

Unshakable Faith, Booker T. Washington and George Washington Carver, John Perry, Multnomah Publishers, Inc., 1999

Sex and God at Yale: *Porn, Political Correctness, and a good Education Gone Bad*, by Nathan Harden, Hillsdale College Newsletter, 2018

Unbroken-*A World War II Story of Survival, Resilience, and Redemption*, by Laura Hillenbrand, Random House, New York, NY, 2014

Darwin's Plantation, by Ken Ham and Charles Ware, with Todd A. Hilliard, Master Books, Green Forest, AR, 2007

The New Politics of Sex: *The Sexual Revolution, Civil Liberties, and the Growth of Governmental Power*, by Stephen Baskerville, Angelico Press, 2017

The Case for Christ, by Lee Strobel, Zondervan, 1998

Free to Choose, by Milton and Rose Friedman, Harcourt, Inc., New York, NY, 1980

The Biblical Basis for Modern Science, by Henry M. Morris, Master Books, Green Forest, AR, 2002

Lee-The Last Years, by Charles Bracelen Flood, Houghton Mifflin Company, Boston, Mass., 1981

Killer Angel (A Biography of Planned Parenthood's Margaret Sanger), by Georg Grant, Standfast Books and Press, Inc., Franklin, TN, 2014

War Against the Weak, by Edwin Black, Four Walls Eight Windows, New York, NY, 2003

Tower of Babel, by Bodie Hodge, Master Books, Green Forest, AR, 2013

The Genius of Ancient Man, by Don Landis, (gen. editor) and Jackson Hole Bible College, Master Books, Green Forest, AR, 2012

The Intimidation Game: *How the Left is Silencing Free Speech*, Kimberly Strassel, Twelve Hachette Group, Inc., New York, 2016

Slaughter of the Dissidents, by Jerry Bergman, Leafcutter Press, Southworth, WA, 2008

Inherently Wind-*A Hollywood History of the Scopes Trial*, DVD, Dr. David Menton, Answers in Genesis, Hebron, KY, 2003 (Points out the inaccuracies of the movie *Inherit the Wind*)

Dupes-*How America's Adversaries Have Manipulated Progressives for a Century*, Paul Kregor, ISI Books, 2010

Expelled: *No Intelligence Allowed*, DVD, DVD, by Ben Stein, 2008

Darwin's Deadly Legacy, DVD, hosted by D. James Kennedy, PhD, Coral Ridge Ministries, Ft. Lauderdale, FL, 2006

Godless, by Ann Coulter, Random House, 2006

Trilogy, by Francis A. Schaffer, Crossway Books, Wheaton, IL, 1990

Lee's Real Plan at Gettysburg, by Troy D. Harman, Stackpole Books, 2003

Stonewalled: My Fight for Truth Against the Forces of Obstruction, Intimidation, and Harassment in Obama's Washington, by Sharyl Attkisson, Harper Paperbacks, New York, 2015

APPENDIX C

Mini-Glossary, and a few Heroes

Abubakar Abdullahi-courageous Nigerian Muslim imam who sheltered several hundred Christian farmers inside his home and mosque from murderous attacks by rampaging fanatic Muslim herdsmen (June, 2018), even offering up his own life for their safety. Reports indicate at least 80 Christians were indeed murdered around his village that day and many hundreds more during that same time period elsewhere in Nigeria, but at least three times that many were saved by his heroic actions. The State Department of the U. S. formally recognized the imam's heroic actions, and awarded him the 2019 International Religious Freedom Award.

adaptations-minor biological changes, or variations, in organisms based on the already-existing genetic information (no new genetic input; *never* leads to large-scale or "macro"-evolutionary changes)

antediluvian-belonging to the time before the biblical Flood, approximately 4004 B.C. to about 2248 B.C., or some 4,300 years ago and before, back to the Creation date

Anthony Joseph ("A. J.") Foyt, Jr.-one of the toughest, most versatile, and all-around greatest auto racing drivers in American motorsports history. Now 84 and married to Lucy Zarr for 63 years, Mr. Foyt's "never-give-up" attitude is legendary, and he would possibly be a first ballot shoo-in for my upcoming all-time "toxic masculinity" All Star team. His auto racing achievements will probably never be equaled, and his many victories include the Indianapolis 500 (4 times) in Indy car; the Daytona 500 in Nascar; the 24 Hours of Daytona (sports cars); and the 24 Hours of Le Mans (sports cars). Mr. Foyt amazingly competed in 35 straight Indy 500s as a driver, despite numerous serious racing injuries, and he won 14 national titles and 172 major races in his stellar driving career, which spanned four decades, three continents, and five countries.

American Civil Liberties Union (ACLU)-non-profit organization founded in 1920 supposedly to defend and preserve individual rights and liberties, but with a strong left-leaning agenda and secular worldview. It works primarily through litigation and lobbying, especially in support of policy decisions established by its board of directors. These include longtime support of same-sex marriage, the right of LGBT people to adopt, opposition to the death penalty, full abortion rights, extreme rights for prisoners, opposition to government and citizen preference for religion over atheism, and rabid support for the teaching of evolution—and *only* evolution in public schools. The ACLU's early signal event relative to evolution teaching and worldview promotion was their extremely successful publicity stunt with Clarence Darrow and the 1925 Scopes Monkey Trial. Americans who embrace Christian family values and a faith-based worldview often find themselves up *against* the ACLU, who columnist Cal Thomas once said "has conducted a religious lobotomy on this country, seeking to strip it of any vestige of religious influence."

apologia-Greek word translated "answer" in 1 Peter 3:15, from which is derived the word "apologetics," which means the careful, logical defense of the Christian faith; somewhat akin to a premeditated, logical *legal defense* in a courtroom setting

"bless her (or his) heart"-a short, often introductory phrase usually uttered with a soft, seemingly sympathetic tone of voice and gentle facial expression, often used particularly in the southern United States, which allows the speaker to subsequently point out poor wardrobe selections, personality flaws, eating disorders and miscellaneous poor lifestyle choices of a particular individual, sometimes scathingly so, but in a socially acceptable manner.

Bob Feller (1918-2010)-American professional baseball and military hero who pitched for the Cleveland Indians for eighteen years. Mr. Feller, known as "Bullet Bob," was an eight-time All Star, seven-time strikeout leader, and was at times unhittable while amassing a win-loss record of 266-162, 279 complete games, 44 shutouts and three no-hitters. His stellar pitching career was interrupted while in his prime by four years of naval service in WW2, when he volunteered for the U. S. Navy right after visiting his terminally-ill father and hearing about the Japanese surprise attack on Pearl Harbor. Mr. Feller was the first American professional athlete to enlist. He had received a military exemption due to his father's failing health, but he wanted to serve in combat missions. He served as Chief Petty Officer on the U.S. Alabama in mostly the Pacific theatre, was decorated with six campaign ribbons and eight battle stars, and was later made an honorary member of the Green Berets.

Booker T. Washington: African-American Christian leader, orator, educator, and hero who was born in a slave hut in 1856, but overcame numerous obstacles to educate himself and become the first teacher, developer, and President of Tuskegee Normal and Industrial Institute. He was the most influential voice for black Americans and descendants of slaves from the early 1890s to 1915. His no-nonsense emphasis on education and hard work is represented in the following statement: "We're not going to ask our former masters for a seat at the table; we're going to *make* the table" (and later sit at it). He was the first African-American to be invited as a guest at the White House (by Republican President Theodore Roosevelt).

"dinos"-abbreviated form of dinosaurs

"elephant in the living room"-an uncomfortable and challenging problem, opposing evidence, or awkward situation that is difficult to ignore, but that people do not want to discuss or deal with (and try to ignore)

epigenetics-the study of changes in organisms caused by modification of gene expression rather than change or alteration of the genetic code itself

fear mongering-recently revived and effective political, psychological, and social strategy nearly perfected by the political left, of claiming there is a crisis about something—often stridently—when there is none

haughtiness-among human beings, a manifestation of the *status disease* or an otherwise concealed low self- image, with the body language, actions, subtle gestures (including eyebrow movements), as well as speech, words, and manner of speaking with an attitude of being arrogantly superior, or of higher socio-economic class, relative to other divinely created *image-bearers*

Islamic terrorism- 1) terrorism by those of the Islamic faith, which has steadily grown into a serious worldwide problem since the 1972 massacre of Israeli athletes at the Munich Olympics, or 2) *workplace violence*, a term commonly employed during the Barack Obama presidential administration; e.g., the horrific Ft. Hood, Texas shooting spree (Nov. 5, 2009) by disgruntled army psychiatrist and follower of Islam Nidal Hasan, where he fatally shot 13 innocent people and wounded 30 others, while shouting *allah 'akbar*! (Allah is the greatest)

jeet-a common colloquial (but important) question asked on a nearly daily basis (sometimes twice a day), by laid-back residents of the southern U.S., essentially meaning, "Did you eat yet?"

mabbul-Hebrew word for flood, used in Genesis 6:17, and used *only* for the cataclysmic Flood of Noah. There are other words used in the Bible for *local* floods, and Scripture makes it very clear that the *mabbul* covered the tops of all the high hills and mountains (Genesis 7:19-20).

opisthotonus (or -nos) or opisthotonic posturing-bizarre contortions seen in some neurological conditions and epilepsy patients. Also, a particular death pose of animals, especially seen in fossilized dinosaurs found in water-deposited sedimentary rock layers, exhibiting a severe backwards hyperextension of the spinal column, neck, and head—likely indicating gasping for breath as agonizing death approaches. These are sometimes found with fish fossils, which often appear as if snap-frozen in time, writhing in suffocation. Specious alternative explanations for this striking death posture have recently been presented by a few long-ages evolution advocates, in an attempt to deny a recent, violent, catastrophic Flood-borne mud death and burial.

polymath- a person of wide-ranging knowledge or learning, but not always with wisdom

propitiation-appeasement, or reparation, for an injury or crime; atonement

Roger Staubach-U.S. Naval Academy graduate, and former all-pro and Super bowl MVP Dallas Cowboy quarterback great, who was the only Heisman trophy winner to also win an NCAA boxing title. Mr. Staubach is also a very successful businessman, won the NFL Man of the Year Award in 1978, and was awarded the Presidential Medal of Freedom in 2018. He was and is a bona fide hero to many American boys and men.

safe places-areas where sensitive students can retreat to on college campuses, and not be challenged by ideas that are conservative or in opposition to their secular progressive worldview or paradigm, and where they cannot be bothered and bored with facts and truth.

snowflake-modern American term describing a wimpy, easily offended person who tends to proffer unjustified vapid excuses for poor performance, neglect of duty, or irresponsibility and is totally self-focused (but usually doesn't realize it), and has rarely worked hard, sweated or done anything that required vigorous handwashing. Snowflakes struggle to engage or compete substantively in the arena of ideas, and often manifest an entitlement or victimhood mindset. Such individuals have little stamina and resilience, are extremely sensitive, and have *feelings* that are bruised at the slightest touch. Snowflakes cannot tolerate views that oppose their own, but lack the grit to mentally, spiritually, and verbally combat such opposition. They are found in increasing numbers on college campi today.

tannin-Hebrew word translated in the Bible as "dragons," except in Genesis 1:21 as "great whales," where it likely refers to the great marine reptiles, or pelagic dinosaurs, like plesiosaurs

tectonic-relating to the structure and dynamics of the earth's crust, including the powerful and violent internal processes, and those that breach the surface, like earthquakes and volcanism

toledoth-the Hebrew word for *generations*; it was translated in the Septuagint Greek by the Greek word *genesis*, and used in the New Testament only in Matthew 1:1, where it is translated "generation." This is relative to the sections and authorship of the all-important biblical book of Genesis, where one notices these small groups of chapters (or thematic sections) interestingly delineated by the recurring phrase: "These are the generations of . . . ", the repeated introductory statement of each of eleven *colophons*. Ancient narrative tablets nearly always ended in colophons, which had a very precise structure made up of three parts, like the eleven found in Genesis: 1) "this has been the book of . . . " (or "genera-tions of," or "genealogy of"), then 2) the name of the person who wrote the tablet, and 3) a date (or, "the year of so-and-so"). These colophons are essentially *author-ship statements*, formerly referred to as "signatures" by English military officer and archeologist P. J. Wiseman over eighty years ago. So it is "generations"-from the Hebrew word *toledoth*- from which the foundational book of Genesis gets its name. This concept is important in determining authorship of the early sections of Genesis.

tolerance-a word forcibly morphed into something quite different from its original meaning in recent decades by secular-left progressives. It is senseless and sinister at the same time. Not only are you required to put up with-or tolerate-those views dif-ferent from your own, but you are required to *agree with* them, and also adhere to the precept that every lifestyle or worldview is *equally* true and legitimate (except the Christian one), and therefore any criticism or intolerance is unjustified, as there is no absolute truth

Tony Cloninger-pro baseball player from Cherryville, North Carolina who became the only Major League baseball *pitcher* to hit two grand slam home runs in one game

verbicide-the mutilation, or even murder of a word, common in late 20th and early 21st century America. "Men [and women] often commit verbicide because they want to snatch a word as a party banner to appropriate its selling quality" –C. S. Lewis

woke-recently birthed politically-correct word, apparently derived from the African-American vernacular "stay woke", and particularly favored by identity politics aficio-nados and sophomoric social justice 'warriors', supposedly meaning one who is now finally alert to injustices in society-especially racism-with a targeted focus on *white privilege*, the Caucasian majority, and particularly the actions, attitudes, policies, and practices of white males. Use of the term implies the false belief that most people in the U.S. are *not awake*—or *ignorant of*—current or past injustices and are *racist*.

CPSIA information can be obtained
at www.ICGtesting.com
Printed in the USA
BVHW061233030720
582846BV00002B/3

9 781631 292248